Eco-Consciousness in American Culture

Adina Ciugureanu / Eduard Vlad (eds.)

Eco-Consciousness in American Culture

Imperatives in the Age of the Anthropocene

Lausanne - Berlin - Bruxelles - Chennai - New York - Oxford

Library of Congress Cataloging-in-Publication
This book has been requested for registration in the Library of Congress CIP catalog of the Library of Congress.

Bibliographic information published by the Deutsche Nationalbibliothek La Deutsche Nationalbibliothek collects this publication in the Deutsche Nationalbibliografie.
Nationalbibliografie; detailed bibliographic data are available on the Internet at http://dnb.d-nb.de.

Romanian Association for American Studies (RAAS)
Cover design by Andreea Simion.

ISBN 978-3-631-90519-7 (Print)
E-ISBN 978-3-631-91023-8 (E-PDF)
E-ISBN 978-3-631-91024-5 (E-PUB)
10.3726/b21290

© 2023 Peter Lang Group AG, Lausanne
Publicado por Peter Lang GmbH, Berlín, Alemania

info@peterlang.com - www.peterlang.com

All rights reserved.

This publication has been peer-reviewed.

This publication may not be reproduced in whole or in part, nor may it be recorded or transmitted by a retrieval system, in any form or by any means whatsoever, in any form or by any means whatsoever. or transmitted by any information retrieval system, in any form or by any means, mechanical, photochemical, electronic, photocopying, recording or in any form or by any means, whether mechanical, photochemical, electronic, magnetic, electro-optical, magnetic, electro-optical, electro-optical, magnetic magnetic, electro-optical, photocopying, or any other means, without the prior written permission of the publisher.

Contents

Eduard Vlad
Introduction: Eco Imperatives and Their Echoes in American Studies 7

Part I: Eco-Consciousness: Politics, Policies, Theories 19

Philip John Davies
The Eco-System of US Politics 21

Eduard Vlad
The Great Reset and Its Pandemic Narratives 41

Daniel Clinci
Posthumanism and Climate Change: Why Theory Matters 57

Roger L. Nichols
American Indians and Developing Eco-Consciousness 75

Loredana Bercuci
Matter, Objects and Nature Knowledge in William Byrd II's *Histories of the Dividing Line betwixt Virginia and North Carolina* 93

Dragoș Osoianu
Material Ecotheological Implications of Food Eating 105

Part II: Eco-Consciousness and Eco-Activism 119

Adina Ciugureanu
Remapping *The Waste Land* and Climate Change 121

Alina Cojocaru
Transatlantic Perspectives on Urban Environments in T.S. Eliot's *The Waste Land* and William Carlos Williams's *Paterson* 135

Oana-Celia Gheorghiu
Re-Imagining *The Waste Land*: Infertility, Barrenness and Ecocatastrophe in Margaret Atwood's *The Handmaid's Tale* 151

Michaela Praisler
"Is this the house you want to live in?" Fictional Warnings from a
Feminine Eco Consciousness 161

Andreea Cosma
Ecopoetics in Diane di Prima's *Revolutionary Letters* 173

Ludmila Martanovschi
Ecodramaturgy Meets the Arctic: Chantal Bilodeau's *Sila* and *Forward* 185

Part III: Ecocritical Readings 201

Anouk Aerni
"Time kept them there and time would let them leave": An Ecocritical
(Re)Consideration of James Webb's *Fields of Fire* 203

Pauline Boisgerault
"Not the glow of red or green as in picture-book terror wolves, but a
dullish, perversely dignified human gold." An Eco-Critical Reading of
Hold the Dark by William Giraldi 219

Carla Francellini
Ecological Consciousness and Environmental Migration through
Animal and Machine Tropes in John Steinbeck's *The Grapes of Wrath* 233

Olga Thierbach-McLean
Uncanceling the Future: Cyberpunk, Solarpunk, and the Rebellion of
Eco-Optimism 251

Florian Andrei Vlad
Rednecks Gone Wild: Ecodefense and Edward Abbey's Monkeywrenching 267

Patrycja Pichnicka-Triverdi
Ecological Thinking in Fantastic Literature. Symbolism of New Heroes:
A Case Study of 21st Century American Vampire Narrative 283

The Authors 301

Eduard Vlad

Introduction: Eco Imperatives and Their Echoes in American Studies

Apparently, such concepts as *the environment* and *ecology* are to be perceived as "purer" than environmentalism and various forms of ecologism. A long Enlightenment tradition has associated what we now call environment studies and ecology with "objective" science, while the associated "isms" show militant, ideologically driven movements. Some people in the "hard sciences" might still think of mathematics as "pure," and of ecofeminism as "impure," political, etc. As a response to this long Enlightenment tradition, there is a growing trend to see many of the concepts which appeared to be "pure" as ideologically tainted. Space is now "tainted by power" in geopolitics, while culture is no longer merely elite, high culture, but also popular culture, treading the "muddy" ground of ordinary daily experience. So it goes with environment and ecology.

A reliable source, such as an *Environmental Encyclopedia*, appears to support such a claim: "Etymologists frequently conclude that, in English usage at least, *environment* is the total of the things or circumstances around an organism – including humans" while ecology "is focused on studying the interactions between *an organism of some kind* and its environment" (Young 467). In his entry in the above-mentioned encyclopedia ("Environment"), Young shows an "impure," ideological bias that most of us share: the environment is what is around humans, so human beings are central, while ecology equally appears (although less obviously) to be anthropocentric. If one can disagree with the second claim, as soon as one adds an -ism to ecology, its ideological, "impure" dimension is made clear, with the more striking *ecoterrorism* as a good example. Equally, when eco is followed by other fields thus acquiring an environmental or ecological turn, the militant, engaged orientation becomes visible: ecodrama or ecodramaturgy is another example. There will be many twists and turns in the current volume, with diversity as its main "ecological" feature.

Among the many turns in the human sciences that have animated scholarship over the last decades, the environmental turn has acquired particular prominence lately. The problematic, stringency, imperatives of ecological and environmental issues justify the publication of the current volume as well. Environmental and ecological concerns feature prominently in the human sciences as a whole, clearly including the literary and cultural studies.

Back in 2014, Sverker Sorlin, while acknowledging the existence of the environmental turn in the human sciences, referred further back to ecologist William Vogt (*The Road to Survival*, 1948), but also to novelist Rachel Carson (*Silent Spring*, 1960). Carson's sinister "elixirs of death" in Chapter 3 drew attention to the devastating effect of pesticides, such as DDT, creating a powerful response among American audiences, raising people's awareness. Since then, an eco-consciousness has been made even more conspicuous due to a perception shared by ever larger groups of people concerning rising atmospheric and water pollution, loss of biodiversity, climate change, both local and global crises of environmental degradation, with far reaching economic, political and geopolitical implications as well (Sorlin 2014).

The issues addressed by the essays in this volume, written by scholars from various parts of the "global academic eco-system," so to speak, are far ranging, while also showing diversity of opinion and approach, thus evincing the diversity of an ever growing interdisciplinary field of environmental and ecological concerns ranging from the natural sciences to the humanities. Are eco-systems harmonious, balanced, stable? As a rule, in the long, biological run, the answer is negative, the dinosaurs bearing testimony to this. Nevertheless, a certain stability is preferable when it comes to humankind's impact on what has come to be called the *Anthropocene*. In environmental and ecologically driven approaches, the term "has emerged as a hot topic of discussion across the full gamut of academic disciplines" Jamie Lorimer notes (Lorimer 117) while promoting, interestingly enough, the alternative spelling of *anthropo-scene*, probably in order to stress the dramaturgic dimension of the environmental show going on in the biosphere, in which humans have assumed center stage position. On the environmental scene in the Anthropocene, humans can choose to play the heroes or the arch villains in the contexts of a growing eco-consciousness. We humans have bitten from the tree of environmental knowledge and what follows is largely attributable to human agency.

If such *environmental stressors* as earthquakes, typhoons, volcanic eruptions, high levels of UV radiation on mountain tops are inevitable, human made pollution, deforestation, desertification, environmental-friendly energy policies can be addressed by concerted action worldwide. If humans exploit the biosphere, they should also be more instrumental in maintaining *ecological integrity*, a key environmental concept, but much more than that, a vital eco imperative.

In interdisciplinary fields like the ones taking shelter under the shadow of contemporary eco imperatives, the concepts and the views expressed by various voices from different backgrounds might need clarification, which is usually facilitated by placing them in identifiable contexts. Otherwise, they might lead to misunderstandings and misinterpretations, to failure of effective communication. In these fields having to do with important issues such as the survival of humankind and of the biosphere on Earth, words and statements have other functions than those ascribed by some to poetic language and its "forest of symbols," as in Baudelaire's famous "Correspondances" in his *Les fleurs du mal*: "La nature est un temple ou de vivants piliers/ Laissent parfois sortir de confuses paroles/ L'homme y passe à travers des forêts de symboles/ Qui l'observent avec des regards familiers (Baudelaire 19). Taking issue with Baudelaire, more prosaic environmentalists from the human sciences might admire the forest of symbols, while paying attention that its theoretical pillars are unambiguous concepts that are to be exchanged in scholarly debates. These prosaic environmentalists might try to make these confusing words clearer, thus facilitating debate at the risk of killing the spirit of poetry, others will claim.

The essays in this volume are largely meant, in addition to addressing stringent environmental and ecological issues, to stimulate debate, to be proved "wrong," whenever possible. Controversies are sometimes conducive to deeper understandings of complex phenomena, benefiting both the "wrong" and the "right" side of an argument, scholarly or otherwise. Let them be read and dealt with as such.

The first environmental concept evoked in the volume is the American political eco-system. It is obvious from its very title that the focus of Philip Davies's "The Eco-System of US Politics" is not the world's great outdoors and the challenges the natural environment and the environmental activists face. However, what happens in what the author calls the American political eco-system is likely to have considerable impact on subsequent environmental policies of a world in which the US is a major player. Philip Davies examines the current eco-system of the American polity, recent developments and possible future trends. At a time when eco-consciousness appears to be growing, in response to multiple threats to the health and equilibrium of the planet and its inhabitants, there is a confrontation embedded in political culture ongoing in the USA that is likely to produce echoes, for the better or for the worse, for a decade or more. While showing some reservations, the author makes pertinent observations with a view to placing eco-consciousness, such issues as climate change and renewable forms of energy in relation to the emerging trends of the eco-system of US politics, and the opportunities liable to promote the nation's ability to convert a growing eco-consciousness into a significant eco-policy. All this is to be seen while acknowledging that an eco-system, including political eco-systems, would involve a great deal of equilibrium, which in politics would involve negotiation, debate, reasonable compromise of the competing "political organisms" of the overall body politic. However, the tendency in American politics over the previous decades, but especially since the success of disrupting forms of populism, such as those promoted by Donald Trump, is to see politics as war, which runs counter to democratic ("eco-friendly") values and practices. Will the American eco-system get better as the next Presidential elections are drawing near, especially if one possible candidate's fervor to win may trump the political card game?

In his "American Indians and Developing an Eco-Consciousness," Roger L. Nichols chronicles developments over the last half century which have shown how the demonstrations conducted by some American Indians against federal and state actions which were threatening their cultural and religious sites have both been supported by, and have also contributed to, other groups involved in awareness-raising concerning ecological issues of widespread interest. The Native Americans militated for their cultural practices and religious rites to be recognized as legitimate, leading to debates, disputes and even clashes with federal agencies overseeing much of the federal land in the West, specifically when federally approved measures affecting the environment interfered with tribal religious sites or rites. When the U.S. tended to dismiss American Indian claims for religious freedom in special consecrated sites, tribal people moved toward the promotion of environmental arguments that turned out to enlist the sympathy and the wide support of a series of organizations, both ecologically aware and religious. This combination went on to show how alliances were made, which otherwise might not have supported tribal demonstrations relying on purely religious grounds only.

The theoretical background of Dragos Osoianu's "Material Ecotheological Implications of Food Eating" harks back to Material Ecocriticism and Ecotheology. The latter, the author claims, integrates both a panentheistic and a pantheistic vision of post-systematic theology. Starting from the symbolic ritual of the holy Communion as consecrated by Christianity, from a material ecotheological point of view, Osoianu thinks, there is a balance of exerting agentive power between the epistemic subject and the ontic object, with the human being consuming the materiality of nature and, at the same time, the environmental others. Apparently, the urge to engage in the holy communion discriminates between the moral categories of good and evil, but the binary opposition does not belong itself to the material world, because nature is good per se in a Christian ecologically minded consciousness.

In "The Great Reset and its Pandemic Narratives," Eduard Vlad starts from conceptualizations and events that became visible in what one might call "the Covid-19 age," or the age of "the pandemic narratives of the Great Reset." The essay places Klaus Schwab and Thierry Malleret's controversial *Covid-19: The*

Great Reset in relation to the various contexts in which it was published, for a better understanding of the outrage that it occasioned, while also giving rise to a host of conspiracy theories as the raw materials and the pins and bolts of what are called here "pandemic narratives." Since a bewildering array of things become viral these days, pandemic-like, rather than network-like, a pandemic narrative is a suitable visualization of discourse displaying extreme connectivity and contagion, but also involving "contradictory pluridirectionality," as it were, often wreaking havoc largely by means of the sensationalist messages disseminated by influential voices in the social media. Contamination, containment, and resistance to it, as well as conspiratorial elite circles will nevertheless feature prominently in the Covid-19 age, which explains why pandemic narratives and conspiracy theories help each other invade the many global pathways of information, misinformation, disinformation. In the opinion of Eduard Vlad, the host of conspiracy theories imagined by influential social media figures as diverse as J.P. Sears and Russell Brand is largely due to a certain (mis) understanding of a famous Great Reset slogan, which will be duly contextualized in the essay.

In "Posthumanism and Climate Change: Why Theory Matters," Daniel Clinci starts from the basic assumption that a theory for what some environmentalists call the Anthropocene must consider and examine climate change, but also engage critically the whole underlying system of global domination and inequalities. A theory for the Anthropocene is very much necessary, as it may provide a framework not only from the points of view of philosophy or ontology, but also from those of politics, society, and economics. In this sense, Clinci notes, theory matters, even if it seems to be currently exiled, in his opinion, to the margins of both academia and activism. A note of reservation and hesitation is, however, voiced, as the author wonders whether pessimistic theories will be confirmed and societies will collapse under the weight of an unjust system, or they will reframe and re-construct the trans-species communities that we have always been functioning in. Critical theory matters as the indispensable tool in the process of re-thinking the world in terms of dealing with the anthropogenic climate change and attending significant human impacts on the planet's eco-systems from an engaged, involved, active perspective with posthumanism as one of the newly emerged approaches to the Anthropocene.

In her essay, "Matter, Objects and Nature Knowledge in William Byrd II's Histories of the *Dividing Line betwixt Virginia and North Carolina*," Loredana Bercuci goes back well beyond the time communities became distinctly eco-conscious, to early 18th century colonial America. William Byrd II's text and its attending conceptions of the self and of nature are worth examining in terms of their historicity and the light the document sheds on one particular episode in the development of what would be eco-consciousness as more than a worldview at individual and group level toward distinct policies at higher ranks of social and political levels. Bercuci is intent on investigating, through her own eco-conscious, material ecocritical perspective, the intersections between environmental knowledge, race, and objects in order to show how the objectification of the boundary line works to objectify non-white races, as well as nature, and to conceal whiteness in colonial America.

Adina Ciugureanu's "Remapping *The Waste Land* and Climate Change" is designed to re-visit T.S. Eliot's major poem through both ecocritical and geocritical perspectives. Ciugureanu argues that the combination of the two perspectives – reading the poem as a prophecy of the possible apocalyptic death of the earth by lack of water and as a geographical mapping of London in the interbellum period – reveals both Eliot's modernist reading of urban space and his knowledge of the climate change views of the time. Re-mapping Eliot's city, the essay discloses and explores the significance of space and place in the poet's cultural, but also geographical reading of the world. Eliot is shown to create both a textual geography and a literary cartography of Western civilization in the aftermath of World War I. On the other hand, space is known to be configured by environmental markers, which could be private, public, general, or even climate zones, such as deserts. The exploration of these markers from the perspective of the increasing volatility they are threatened by, of crises in the social and natural worlds they represent, reflects an ecocritical concern. A joint ecocritical and geocritical approach offers lines of inquiry that examine the intersections between mapping literary spaces and exploring ecological consciousness. This is the line of reading which is followed by the revisitation of *The Waste Land* through remapping its spaces and reconsidering the apocalyptic scenes which the poem reveals.

In "Transatlantic Perspectives on Urban Environments in T.S. Eliot's *The Waste Land* and William Carlos Williams's *Paterson*," Alina Cojocaru compares and contrasts two apparently distinct perspective on the Modernist urban space grounded on either side of the Atlantic. The urban environments in T.S. Eliot's *The Waste Land* and William Carlos Williams's *Paterson* are reassessed from a contemporary perspective marked by the shift from an anthropocentric perspective towards an ecocentric perspective. The two great Modernist poems may be separated by over two decades, yet they are arguably complementary in their perspectives on the urban environment and the "urban climates" concerning the new ideas, philosophies and politics that permeate the city on the two continents following the First World War and the Second World War, respectively. These major poetic works by T.S. Eliot and William Carlos Williams, although rooted in the same Imagist theory of Ezra Pound, express the tensions between the high-brow and low-brow modernism of the 20th century. If the former is attuned to the classical tradition, William Carlos Williams responds to the Anglocentric cosmopolitanism of *The Waste Land* by dwelling on the margins of traditions and grounding *Paterson* in American realities.

Oana-Celia Gheorghiu is another contributor whose essay, "Re-Imagining *The Waste Land*: Infertility, Barrenness and Ecocatastrophe in Margaret Atwood's *The Handmaid's Tale*," probes the creative surfaces and depths of T.S. Eliot's major Modernist poem. She uses the "chessboard" of T. S. Eliot's *The Waste Land* as an intertextual reference point for the examination and interpretation of Margaret Atwood's *The Handmaid's Tale* with a view to highlighting the *wasteland* of womanhood suppression and women's rights abolishment following an environmental and political catastrophe. This is prompted from the title and text of Eliot's second section of the poem, "A Game of Chess." *The Waste Land* is the perfect example of a puzzle, which has not yet been unequivocally glued together, and which is, in the opinion of Gheorghiu, still awaiting creative or insightful solutions for sorting out its apparently intended disorder. The poem – with its *heap of broken images* (I, 22), resulting from the central theme and its representation: the world's barrenness in the wake of an apocalyptic Great War which left marks on human consciousness – is paired in Gheorghiu's "chess game" with a "novelistic Queen" which turns out to be Margaret Atwood's *The Handmaid's Tale*. Atwood's work provides clues for the interpretation of some of that Eliotesque heap of broken images that the author of the essays intends to arrange and to make sense of. In so doing, Gheorghiu justifies the title of her essay.

Michaela Praisler's "'Is this the house you want to live in?' Fictional Warnings from a Feminine Eco Consciousness," while leaving Eliot's *The Waste Land* behind, also leaps into the future, examining the dystopian world of Atwood's *The Handmaid's Tale*, while the title invites us all to answer in the negative. Margaret Atwood's novel highlights environmental wrongs and advocates through its fictional rhetoric women's rights. It evokes dystopian patriarchal times and creates a literary puzzle through clues woven into its fabric and through the closing paratext. In touring Atwood's house of fiction, Praisler's chapter also considers the author's relevant non-fictional interventions – interviews, lectures and speeches with a view to supporting her ecofeminist argument. The rise of dystopian, catastrophic literature has supported the rise of eco-consciousness and eco-criticism, with the corresponding environmental activism. Among the many branches of eco-criticism that are seen as starting points for contemporary debates, particularly urgent, in the opinion of the author of the essay, is that which brings together environmental wrongs and women's rights: ecofeminism. The final fictional warning on the world and the word, on the erasure of topics related to women and the environment that Atwood formulates in her fictional work and other public may not have reached a very large (or very enthusiastic) public. Her decision to engage in open debates on women's rights and in environmental activism via multiple media, thus also addressing numerous others besides the avid traditional literature readers is highly commendable. Praisler expresses her hope that the academia will leave their ivory tower to visit the old house society is inhabiting – whose layout still reveals upper male exclusivist clubs and lower female servants' quarters, but which is now, undergoing dramatic threats from specific natural disasters and far-ranging ecological disruptions.

How do ecodramaturgy and the Arctic meet? To find the answer to this question, one is to read Ludmila Martanovschi's essay, "Eco drama Meets the Arctic: Chantal Bilodeau's *Sila* and *Forward*." *Sila* (2015) and *Forward* (2017) are the first two plays in Chantal Bilodeau's ecodrama series exposing climate change in the countries of the Arctic Circle, highlighting essential themes for ecodramaturgy. In the two plays, the writer-activist explores gendered responses to the environmental crisis over several decades, reflecting on the artist's role in helping audiences visualize it. *Sila* and *Forward* illuminate, in the opinion of the author of the essay, the most relevant themes and techniques of ecodramaturgy. They raise awareness about current ecological issues in memorable ways, while their subject matter is distinct: *Sila* features an Inuit mother who loses her adolescent son in the context of contemporary pressures placed on their indigenous community, the human story paralleling that of a female polar bear and her cub, while *Forward* looks at the ecodramatic story of icescapes, an Arctic explorer and his complicated legacy.

Anouk Aerni's essay's title is "'Time kept them there and time would let them leave': An Ecocritical (Re)Consideration of James Webb's *Fields of Fire*." The author of the essay sets out to demonstrate the ways in which James Webb's 1978 novel challenges the absolutism of the binary opposition of humans and nature and instead suggests a relation based on interconnection and interaction. Through its use of particular themes and devices – including but not limited to the decentering of the human, fragmentation, anthropomorphisms and the omnipresence of death – the novel rejects anthropocentrism as well as the ideas of human exceptionalism and anthropological difference. this paper will show that reading postwar literature ecocritically may generate vital contributions to an ecologically motivated reformulation of the complex relationship between nature and humans. The author of the essay examines the ways in which Webb's novel's overall fragmented structure and the fragmentation of its characters decenter the human and thereby challenge the fundamental axioms of Western anthropocentrism. Focus then shifts towards the tension between empirical and experiential time, showing how this will further distance the human characters from their cultural background and how anthropomorphisms and amalgamations of the human and the nonhuman contribute to an establishment of a more nuanced understanding of the nature-human relationship, how all of these aspects work together to paint nature, humans and their relationship in a novel written by a Vietnam War veteran.

Florian Andrei Vlad's essay is a contribution to the examination and contextualization of one of the best-known illustrations of anarcho-environmentalism or radical environmentalism which made Edward Abbey an iconic figure. The title shows one group of militants, one environmentalist approach and one particular practice: "Rednecks Gone Wild: Ecodefense and Edward Abbey's Monkeywrenching." The essay invites one to see the author exploring the "eco imperatives," gradually moving away from anthropocentrist to biocentrist positions that have gained prominence in ecocritical studies over the last few decades, with touches of anarchism announcing his monkey wrenching approach in the novel that this essay focuses upon. The militant dimension evoked by the monkey wrench is to be linked to the almost equally "luddite" term of "ecodefense." Abbey's fictional account of his eco-terrorist gang does not appear to spread fear, the "revolutionary" behavior of what may appear to be a group of rednecks having fun in the great outdoors of the American Southwest being a far cry from the terrible deeds of the Glanton gang in the same area one century before in Cormac McCarthy's *Blood Meridian*. Abbey's characters roam the environment of the desert, moving in the ideological realms between fake terrorism and eco-terrorism, never settling down to one particular position, between the rough and tough anarchist eco-warriors and the reflective, pensive romantic rednecks in love with the great outdoors of the sun-drenched stony Southwest. Deadly serious at times, mock-heroic and parodic at times, engaging with the conventions of the picaresque improved by those of the western, Edward Abbey's ecodefense, although a period piece of the early days of the radical eco- imperative movements, can be re-read now from the perspectives forced upon the world by the somehow unexpected developments brought about by the apocalyptic pages of A.D. 2022.

At first sight, Olga Thierbach-McLean's perspective is far from one in keeping with most kinds of dystopian speculative fiction, judging from the title of her essay, "Uncanceling the Future: Cyberpunk, Solarpunk, and the Rebellion of Eco-Optimism." Thierbach-McLean notes that in the face of present-day environmental challenges, it often seems like we have inherited a present without a future. This eco-melancholic mood is considered by the author to be a product of decades-long cultural autosuggestion. The dystopian forecast of *cyberpunk* fiction has conditioned audiences to think of humanity's prospects as inescapably bleak. The genre of *solarpunk* is emerging in counter-reaction to this trend. Refusing to give up on the future, it strategically cultivates optimism, while also looking for ways of stimulating radical ecological reform. Early cyberpunk works such as William Gibson's genre-defining novel *Neuromancer* (1984) set the tone by describing corporate dystopias in which the capitalist marketplace has become the only source of meaning and value. Recent cyberpunk-themed releases like Steven Spielberg's *Ready Player One* (2018) or Robert Rodriguez' *Alita: Battle Angel* (2019) keep reproducing classic genre tropes such as corporate greed, unrestrained consumerism and environmental destruction while all but abandoning the socio-critical aspects of early cyberpunk. The current essay goes on to explore the budding genre of *solarpunk*. Often defining itself in direct opposition to cyberpunk features – setting hopefulness against cynicism, communal cooperation against individual struggle, lush art nouveau aesthetic against techno-orientalist neon-noir –, its mission statement is to explore paths to a desirable future by formulating possible solutions to avoid environmental destruction.

Pauline Boisgerault undertakes, as the title of her essay shows, to view through an eco-critical lens the novel that Jeremy Saulnier turned into a thriller. The same title also shows that Giraldi's novel is equally meant to thrill and chill, in the wake of the American Gothic tradition, if only by the power of words on the page: "'Not the glow of red or green as in picture-book terror wolves, but a dullish, perversely dignified human gold': An Eco-Critical Reading of *Hold the Dark* by William Giraldi." Estranged from their family or community, the novel's characters struggle to regain a sense of place in their environment. They can only hope to survive by acknowledging their fault towards the many wolves of the novel, both hybrid and animal figures. After a series of murders perpetrated on children, for which the wolves are falsely accused, the hunt through the uncanny Alaskan landscape sheds its gloomy light on its contaminative and transgressive nature. Yet, transgressions allow the novel to offer a de-territorializing perspective on human and non-human relations, based on a blurring of boundaries.

Carla Francellini revisits and reassesses one of Steinbeck's pillars to fame, his *Grapes of Wrath* within the environmental coordinates that the present volume observes. The title of the essay is "Ecological Consciousness and Environmental Migration through Animal and Machine Tropes in John Steinbeck's *The Grapes of Wrath*." Francellini considers that the novel takes a two-pronged approach to exploring the causes of environmental destruction and the massive Dust Bowl migration of the 1930s, representing in fictional form all too real economic gain and class conflict issues. While questioning the impact of excessive mechanized farming, Steinbeck points to the tenants' excessive plowing, which caused soil damage. The novel suggests that scientific progress must prioritize the well-being of humans and the environment, emphasizing the importance of a healthy relationship with Nature for the benefit of all. John Steinbeck dramatizes the link between economic crisis and climatic displacement, depicting environmental migration as a result of both, while emphasizing how human actions and technology can damage the environment and the land on a smaller scale. The writer's use of animal and machine symbolism is crucial to the novel's structure and meaning, as it connects the Joads' journey with the chapters that give the story universal significance. Animals and machines are noted to play a significant role in conveying Steinbeck's most arresting symbolic and allegorical meanings.

Patrycja Pichnicka-Trivedi considers that vampire narratives are rich in ecological meanings. She explores the ecological significance of some of them in her essay, "Ecological Thinking in Fantastic Literature. Symbolism of New Heroes: A Case Study of 21st century American Vampire Narrative." These narratives tend to revolve around the representation of the attitude towards non-human beings (the definition and application of such notions as agency, subjectivity and personhood) and the practical relations between man and nature. In opposition to the "bad" vampire, the new vampire figure is not evil and frequently appears as vegetarian. The essay examines the ecological significations in the new Vampire narratives and their relation to posthumanist challenges and prevailing anthropocentric views. Among the ones the author examines is the ambivalent tragic vampire in Anne Rice's *Interview with the Vampire*, seen as one of "the most ecological," a disturbing figure of the (post)modern condition. A (self) destructive desire was critically exposed, yet the non-human vampire was not condemned. In 21st century vampire narratives Pichnicka- Trivedi finds such vampire heroes as in *Hemlock Grove,* an heir and a product of medical industry, fighting his nature by finally destroying all that he loves in the consuming frenzy, and Abby in Matt Reeves *Let Me In* (2010), another ambivalent vampire, a product of the internal hidden violence of contemporary society. Their nature is defined by the transgression of humanity – which makes humanity a permanent point of reference in a contested nature – human interaction.

"Ecopoetics in Diane di Prima's *Revolutionary Letters,*" by Andreea Cosma, is concerned with the representation of space in di Prima's 1968 *Revolutionary Letters*. The essay engages with the analysis of real, imagined and ideological spaces, described as what Cosma calls "metamanifest places," fostering a sense of social and environmental progress combined with aversion towards forms of oppressive authority and ideology still prevailing at that time of countercultural change. The *Letters* are seen as a literary manifesto against prevailing capitalist discourse, positioning women as ideological and environmental activists. Diane di Prima tackles the topic of the environment from both an aesthetic and a political view, the volume showing her author as both artist and activist. Di Prima's ecopoetics can be seen as convincingly representing the surrounding space, as well as people and places that were meaningful to the poet and that led to her preoccupation with environmental protection. The poetic voice in the *Revolutionary Letters* seems to be fully engaged in the described experiences, turning di Prima's poems not only into political statements, but also into authentic artistic creation devoted to the natural environment. The environment, then, is both natural and artistic, and both are culturally constructed by the power of words in spaces which combine their physical concreteness with artistic configuration and social and political practice, reminiscent of Edward Soja's influential "trialectics of space."

Questions as to whether one witnesses the development or the emergence of a "post-truth" political eco-system in the wake of important events in 2016 and in the (post) Covid-19 age, in which pandemic narratives rose to contaminate the more comprehensive cultural and geopolitical eco-systems are open to debate. So are those concerning the ecotheological implications of food eating or the links between the development of an eco-consciousness and the preservation of Native American religious rites or about finalizing a theory of the Anthropocene as essential for solving the serious problem of climate change. How rich are vampire narratives in ecological meanings? The questioning might continue, thus re-evoking some of the issues that the contributors to this volume raise, examine or deal with. Careful, critical readers are able to read and engage with what these contributors have to say, irrespective of the possible misreadings that these introductory lines might display, in a comprehensive, general, global attempt at finding ways to take eco imperatives in *dead* earnest. Before it is too late.

References

Baudelaire, Charles. *Les fleurs du mal*. Paris: Poulet – Malassis et de Broise, 1857. Print.

Carson, Rachel. *Silent Spring*. Boston: Houghton Mifflin, 1962. Print.

Sorlin, Sverker. "Environmental Turn in the Human Sciences: Will It Become Decisive Enough?" Institute for Advanced Study, 2014 Environmental Turn in the Human Sciences – Ideas | Institute for Advanced Study (ias.edu) Web.

Young, Gerald L. "Environment." *Environmental Encyclopedia*. Third Edition. Vol. 1 A – M. Ed. Marci Bortman, Detroit/ New York/ San Diego/ San Francisco/ Cleveland/ New Haven, Conn./ Waterville, Maine/ London/ Munich: Thomson Gale, 2003: 467–469. Web.

Part I: Eco-Consciousness: Politics, Policies, Theories

Philip John Davies

The Eco-System of US Politics

Abstract: Responding to the theme of this collection this essay examines the current eco-system of the American polity, its recent development and its possible future course. At a time when eco-consciousness appears to be growing, in response to multiple threats to the health and equilibrium of the planet and its inhabitants, there is a confrontation embedded in political culture ongoing in the USA that is likely to produce echoes for a decade or more. The eco-system of US politics that emerges from this will help determine the nation's ability to convert its eco-consciousness into eco-policy.

Introduction: There may be no natural equilibrium

I doubt that William Riker ever used the language of "eco-systems," but his 1982 book, *Liberalism Against Populism*, seems to me to contemplate themes that concern this topic. Riker considered the populist idea that elections can provide a direct and reliable link, converting the public will into public policy, to be unsustainable. This is especially so if there are diverse opinions held in the electorate, rather than a simple binary division.

Therefore:

- Alterations in the system through which the electoral opinions are determined might produce different political outcomes even while the electorate's decision does not change
- Groups and their allies can form coalitions that create stability from this potential maelstrom, and may sustain that stability for a period
- There is no natural equilibrium towards which the polity defaults
- New actors or newly powerful actors may impact on the polity at any time to change or even abolish equilibrium
- Such change can be lethal

According to the elections scholar Walter Dean Burnham, speaking on the topic of *Critical Elections Revisited* in 2006, "Riker had nightmares about the potential that existed as a result" of his conclusions. [1] (Burnham web).

---

1 Critical Realignments Revisited (utexas.edu)

In 1994 the first mid-term election of the Clinton administration saw control of both chambers of the US Congress switch to the Republicans. Large swings – 54 House seats; 8 states in the Senate – prompted Burnham to suggest this was possibly a major change point in US politics, but he did not discern an emerging, stable, new equilibrium. Rather, he predicted "an impending crisis moving towards some kind of upheaval" (Burnham web).

From a "*Broken Branch*" to "*Even Worse*"

Burnham was not the only commentator concerned about America's political eco-system. In 2006 Thomas Mann, of the Brookings Institution, and Norman Ornstein, of the American Enterprise Institute, co-authored, *The Broken Branch: How the Congress is Failing America and How to Get It Back on Track*.

While they identified this breakdown beginning in the latter years of Democratic dominance in Congress, they claim that problems accelerated after 1994. They reserve their sharpest criticism for the Republicans. In this article, published by the Brookings Institution, the authors concluded that the Republican Speaker, Newt Gingrich, appeared initially to favour bipartisan outreach, but allowed this approach quickly to be replaced "by a widespread acceptance by congressional leaders that the ends justify the means," and an "indifference to the importance of the history of the House."[2]

The Broken Branch identifies a growing polarisation of parties; an increase in manipulative practices like partisan redistricting to gain advantage, and a reduction in the willingness to compromise across partisan boundaries. They observed that the practice whereby cross-partisan coalitions supported nationally beneficial legislation was wilting in the face of the desire to win inter-party battles at any cost.

As early as 1993 Eric Uslaner had demonstrated a grasp of this trend, and of its threat, in his book *The Decline of Comity in Congress*. At that time, I didn't understand why he thought it was so important. I took my lead from the 1950 Report by the American Political Science Association's Committee on Political Parties, "Toward a More Responsible Two-Party System." Greater policy consistency and a closer relationship of administration agendas to campaign platforms seemed desirable to me and to fellow political scientists in The Committee for Party Renewal.

---

[2] The Broken Branch: How Congress Is Failing America and How to Get It Back on Track (brookings.edu)

In 1988 the CPR expressed thought that "the national parties have been revitalizing themselves. It is time to begin party building efforts in the states and return state parties to positions of primacy in policy-making" (Committee for Party Renewal 641). However insofar as American political parties have become stronger, this shift has not produced the results that I expected. There is evidently an endemic problem with the creation of a Responsible Two-Party System if either of the parties chooses not to accept the burden of acting responsibly.

Mann and Ornstein were made anxious by what they saw as abuse of the traditional practices of American government to gain and maintain party power. They also noted that, following the Reagan administration's elimination of the Federal Communications Commission Fairness Doctrine in 1987, new media outlets emerged, catering to extreme political audiences. It became easier to foster the fringes of US political opinion to win primary elections, where generally more activist, more opinionated voters take part. Thomas Mann and Norman Ornstein had the intellectual weight and "inside the Beltway" clout to engage some Washington officeholders in their discussions regarding possible reform. The limited success of these efforts may be inferred from the titles of their next two co-authored books, *It's Even Worse Than It Looks* (2012) and the 2016 volume, *It's Even Worse Than It Was*.

"The fervor to win ... trump[s] everything else"

Concern was felt in communities other than those of journalists and academics. In a March 2012 *Harvard Business Review* article, "Fixing What's Wrong with US Politics,"[3] David Moss responded to a contemporary survey of Harvard Business School alumni. Sixty per cent of respondents agreed that the effectiveness of the [US] political system was worse than in other advanced economies and was a threat to US competitiveness.

Moss points out that in previous periods of sharp ideological difference in US history legislative output could nevertheless be productive. Money, media and malign political actions like gerrymandering have a history in US elections but, Moss speculates, these have changed in character, and in combination contribute to a newly problematical political landscape. "The real problem with American politics," Moss reports on behalf of his business colleagues, "is the growing tendency among politicians to pursue victory above all else – to treat politics as war – which runs counter to basic democratic values and may be crippling

---

3   See https://hbr.org/2012/03/fixing-whats-wrong-with-us-politics

Washington's ability to reach solutions that capture the smartest thinking of both camps." He continues, with a prescient choice of words, "The fervor to win too often appears to trump everything else – including respect for opponents, the integrity of institutions, and even the health of the democracy itself" (Moss web).

Making specific reference to the business consequences of the evolving political environment Moss noted the debt-ceiling crisis of July 2011. Republican legislators threatened to precipitate a US government default rather than negotiate with the Obama administration. Default was avoided, but the Standard & Poor credit rating for the USA was downgraded for the first time in its history, reflecting the S&P "view that the effectiveness, stability, and predictability of American policymaking and political institutions have weakened…" and its alarm over political brinksmanship. (Moss web)

Republicans again threatened a debt ceiling crisis in 2013 in an attempt to derail Obama's Affordable Care Act. A partial close down of federal government put 800,000 employees on temporary leave, again prompting international authorities to downgrade the country's credit rating. The debt ceiling resurfaced as battle line in Washington's political warfare in mid-2023, and appears likely to re-emerge again when Congressional chambers and the Executive are held by different parties and one or other perceives a strategic advantage in the confrontation.

In 2017 Mann and Ornstein, with co-author E.J. Dionne, proposed in their next publication, *One Nation After Trump*, a roadmap to take the nation beyond the dangers of the Trump administration. Paralleling some of David Moss's earlier thoughts their recommendations included:

- political elites should engage more with the electorate's anxieties on such issues as jobs and the economy (chapter "With Opportunity and Justice for All: Building a New Economy")
- democratic citizenship should be taught in schools (chapter "Our Little Platoons: The Urgency of a New Civil Society")
- voting should be encouraged and possibly even become mandatory (chapter "What 'Draining the Swamp Really Looks Like: Bringing a New Democracy to Life")
- fake news should be removed from public discourse (chapter "When the Truth Doesn't Matter: The Crisis of the Media and the Rise of Alternative Facts")
- the institutions of civil society should be reinvigorated (chapter "Our Little Platoons: The Urgency of a New Civil Society")

Reviewing in *The Financial Times*, Edward Luce judged these authors' proposals "improbable" and "an Asterix-like magic potion." (Luce web)

The unique era of divided government

Usually, since 1789, the US House and Senate and the Presidency have at any one time been controlled by a single party. In the 20th century up to 1966 this was the case for over 82 per cent of the time. Divided party government was the exception, brief and often associated with shifts in the political parties' electoral coalitions.

From 1968 to the present the pattern has been reversed. Divided party government has become the norm, being present over 70 per cent of the time. This is a unique era in the United States history. Since the late 1960s neither political party has proven capable of building an electoral coalition to provide stable, unified party of national government. Observing this in the late 1990s, I concluded that elections in the USA would continue to be very close, party control of US government may continue to be divided, with the likely result being very combative, probably worse tempered, more media-driven and expensive elections.

I acknowledged the emergence of new media but did not anticipate their huge impact. I did not foresee the rise of a Donald Trump figure, even though Trump had by that time already speculated about running for president. Nor did I recognise how the emerging conditions could clear the way for a populist president of his ilk. I thought it likely that, while it might take a generation or more to happen, there was a demographic inevitability underpinning the eventual emergence of a differently formed, but potentially dominant, Democratic Party. I suggest that the key defining factors *en route* to the current crisis of American democracy are rooted in decisions made by the Republican Party's choice of responses to this possibility.

Leadership, demographics and the political eco-system

Discrimination against African American voters has been long-standing. The Democratic party's domination of the South had a firm foundation in racism, but the party of Lincoln rapidly lost its earlier appetite to engage with African American politics. In 1948 the Democratic Party included in its platform a civil rights plank proposed by Minneapolis Mayor Hubert H Humphrey. The Democratic Governor of South Carolina, Strom Thurmond, bolted the party. His States' Rights campaign for the presidency won in four states.

Black voters still had mountains to climb, but choices had been made. By the mid-1960s Thurmond was a Republican Senator, while a Democratic administration passed the 1964 Civil Rights Act and the 1965 Voting Rights Act. A couple of decades later the South was a firmly Republican region, its politics retaining a racial subtext.

The 1964 presidential election of Lyndon Johnson may have been the last to the present in which a national plurality of white voters supported the Democratic party nominee. The marginal advantage for the Republican party among white voters was small at first, in 1976 being just four per cent. It has expanded since then. In the 21st century the advantage for the Republicans among white voters has fluctuated between 12 per cent in 2008 and a peak of 21 per cent in 2016.

When the Democrats' fabled "blue wall" of Midwest industrial states fell in 2016 to Donald Trump, a common theory was that the white working-class voters of these states had suddenly abandoned their traditional home in the Democratic Party. But research by Nicholas Carnes and Noam Lupu indicates that white working-class voters have been moving gradually to the GOP since the 1990s – a shift over three decades from which Donald Trump benefited, rather than a sudden avalanche that he caused.

White working-class voters have become a more important element of the Republican electoral coalition, but are simultaneously a shrinking portion of the total electorate. They formed a minority even of Trump's voters, who were, on average, quite an affluent group. Nevertheless, the mediated perception that Donald Trump prompted a sharp strengthening of a Republican working-class connection remains strong.

The future is diverse

America's white population is a shrinking part of the voting public. Eighty-eight per cent of voters in the 1980 presidential election were white, and 56 per cent of these voted for Reagan, delivering him 49.3 per cent of the total vote – easily enough to win the election against President Jimmy Carter, Independent John Anderson, and other minor candidates. Donald Trump's 2016 lead among white voters was even larger than Reagan's, at 58 per cent, but now whites made up only 70 per cent of all voters, and whites delivered just under 41 per cent of the total electorate to Donald Trump in the year that he won the presidential election. Trump's victory came in the Electoral College. He lost the popular vote by almost 3 million votes and attracted 2.2 per cent less of the national vote than Hillary Clinton. Trump reached the White House with a campaign tuned closely to the Republican Party's increasing reliance on the white vote, but could not have reached the presidency without the helpful mechanics of the Electoral College.

In 2020 President Trump finished behind Joe Biden by 4.4 per cent, or over seven million popular votes. Nevertheless, the election really was won by a whisker. Biden's 306 to 232 Electoral College advantage would have been overcome if Georgia, Arizona and Wisconsin had been taken by Trump. The Democratic popular vote lead in these states was less than 43,000, or 28 thousandths of one per cent of the national two-party popular vote. A swing of half that would have tied the Electoral College, throwing the choice into the House of Representatives where gerrymandered Republican control of state delegations would almost certainly have seen Trump re-elected.

In 2020 the white vote favoured Donald Trump by 58 per cent to 41 per cent for Joe Biden. In previous elections this might have been the foundation for a solid electoral victory. But the US electorate is becoming more diverse. White voters made up 67 per cent of the participating electorate in 2020, and delivered less than 39 per cent of the total US vote to Donald Trump. Black voters formed 13 per cent of the 2020 vote, splitting 87 to 12 for Biden. Latinos, also 13 per cent of the total, divided 65 to 32 for the Democratic candidate. Asians, 4 per cent of the total, favoured Biden over Trump by 61 to 34. Others, which will include those self-identifying as mixed ethnicity, at 4 per cent, chose Biden by 55 to 41.[4]

For decades, and especially in the 21st century, the Democratic Party has attracted an increasingly diverse electorate, while the Republicans have become increasingly focused on the white voters who form by far the largest ethnic or racial group in that party's electoral coalition. But the white proportion of the electorate continues to decline. According to calculations by Rob Griffin, William Frey, and Ruy Teixera for the Center for American Progress, the Republican Party will cease to be competitive at a national level by 2030 if it fails to do a better job of appealing to minorities (Roy web).

The 2020 Census results showed, for the first time in US history, the population of self-identified white respondents fell not just as a proportion of the population, but also in total number. This reflects the different age and fertility profiles in America's white and non-white population, the ongoing impact of immigration and some change, as mixed race and mixed ethnicity self-identification becomes bureaucratically easier. The continuing trend is towards a population identifying itself in a more diverse manner.

---

4 According to Election Exit Polls 2020 – The New York Times (nytimes.com)

Growth and opportunity

The Republicans in the early 20th century faced the choice of whether to go with diversification, or to build more energetically in the populations already familiar to them. There were voices in the GOP for diversification. Karl Rove, Special Advisor and Deputy Chief of Staff under President George W Bush, claimed in the wake of unusually good Republican mid-term election results in 2002 that people might "look back and say the dam began to break in 2002" (Green web). There had been a noticeable uptick in Hispanic votes for George W. Bush. Rove felt that this growing voting bloc, with its conservative cultural values, firm Catholic religious base, and entrepreneurial inclinations, provided fertile territory for the GOP.

Rove has a point. Latinos are not a homogeneous group. Cuban Americans are often identified as an outlier Republican-supporting group given their experience of the Castro regime, but they are not the only national group who may harbour fears of self-styled left-wing governments. Vocal Republican pursuit of anti-immigration and anti-welfare policies have probably undermined the party's appeal to many Latinos, at least in the early decades of the 21st century. But, historically, not all immigrant groups have automatically become part of the Democrats' electoral coalition and, as their economic and social conditions change and potentially improve, these voting groups' electoral motivations may alter.

In 2008 losing Republican candidate John McCain's gracious words acknowledged the special position held by Obama's election victory: "This is a historic election, and I recognize the significance it has for African Americans and the special pride that must be theirs tonight. We both realize that we have come a long way from the injustices that once stained our nation's reputation."[5] He spoke for many Republicans, proud that their nation had elected an African American to the White House.

This attitude had limits in the GOP. In a 2010 interview Republican Senate Leader Mitch McConnell placed partisan victory above all else, stating that "[t]he single most important thing we want to achieve is for President Obama to be a one-term president" (Hiatt web). Donald Trump and others were meanwhile pressing the claims of the "birther" movement that Barack Obama was not an American citizen. None of this was likely to reach out to African American voters.

---

5   See Transcript Of John McCain's Concession Speech: NPR

After President Obama's re-election in 2012 the Republican National Committee published an extensive autopsy, *The Growth and Opportunity Report*. At its 2013 launch GOP Chair Reince Prebus said that the party needed to adopt a softer tone and broader outreach, include a stronger push for African American, Latino, Asian, women and gay voters. It is not clear that any serious attempt was made to implement its recommendations.

REDMAP, redistricting and system-driven strategies

In seven of the eight most recent presidential elections the Democratic candidate has received more votes than the Republican. The Democrats' problem is that they have not replicated this success in the other branches and levels of the US government. The operation of the Electoral College means that twice, and almost on a third occasion, the party's higher popular vote did not result in gaining the presidency. The structural features that inhibit the conversion of Democratic support into Democratic party officeholding remain in place.

Votes in the Electoral College have never mirrored perfectly the aggregate national popular vote. Each state's Electoral College representation is determined by its number of seats in the US Congress. The number of US House seats is determined according to the states' population figures. The number of US Senate seats is static, at two per state. In the smallest states an Electoral College vote may reflect a population of about 200,000 people, in the largest, approaching 750,000. All but two states deliver their Electoral College votes as a bloc.

The popular vote and the Electoral College vote diverged only three times before the end of the 20th century, in 1824 (when not all states used the popular vote), in 1876 and 1888. But it has happened twice recently, in 2000 and 2016, and very nearly in 2020. The Republican Party's strength in medium sized battleground states improves its chance of winning future presidential elections without the support even of a plurality of voters.

The two Senators per state rule makes the Senate a particularly efficient target for whichever party has strength in the more rural small and medium sized states. Laura Bronner and Nathaniel Rakich calculate that the Senate is currently the most party-biased institution in the US federal government. The GOP in Senate has not represented a majority of Americans since 1999 but has held a majority of seats for more than half of the congressional terms since then. In

2021 their 50 Senators, half of the Senate's complement, represent only 43 per cent of the US population.[6]

The US Supreme Court is created by successive Presidents and their Senates. The Court has had a conservative majority for some time. In recent years Republican Senates led by Mitch McConnell unprecedentedly denied hearings for Obama nominee Merrick Garland for months until the 2016 election passed the presidency to a Republican, but later crashed through the nomination of Trump nominee, Amy Coney Barrett in a few weeks to reinforce the conservative presence before a Democrat returned to the White House. In recent years the Court has stripped away the protections offered by the 1965 Voting Rights Act, undermined the regulation of campaign spending and effectively eliminated the reproductive rights created by *Roe v. Wade*.

In the US House of Representatives constituencies are meant to be equal in population, with the *caveat* that boundaries do not cross state lines. Gerrymandering is the purposeful manipulation of constituency boundaries to advantage one party. A long-established practice in US politics, the application of computing power has made it an increasingly powerful tool. The GOP's rejection of a diversity strategy came at the same time as the launch of the Redistricting Majority Project, or REDMAP, established by the Republican State Leadership Committee, and intended to "erect a Republican firewall through the redistricting process." This was an intensification of a longer-term co-ordinated state-level effort through which "Republicans were able to pack Democrats into a small number of highly blue districts and spread their own voters over a larger number of safe, red districts" (Byler web).

Targeted investment in relatively low-cost state level elections helped produce Republican state legislatures which had the authority to draw new constituency boundaries for state and congressional constituencies after each census. The REDMAP website pointed out the success of this aggressive push on gerrymandering with the example of Ohio, where in 2012 the GOP won 75 per cent of that state's Members of US Congress in spite of gaining 52 per cent of the state-wide vote. Similar successes were put down to the REDMAP operation in Michigan, Pennsylvania and Wisconsin, all part of an attempt to "create 20-25 new Republican Congressional Districts through the redistricting process ... solidifying a Republican House majority."[7]

---

6   See Advantage, GOP | FiveThirtyEight
7   See the opening statement of the REDMAP website: The RSLC Redistricting Majority Project – REDMAP

Each decennial census is followed in many key states by efforts on the part of the dominant party locally to re-design constituency boundaries to their own benefit. The process can be long drawn out, and is not always successful, and may prompt legal action. The current US Supreme Court has signalled its disinterest in taking on these cases, considering such matters as more fit for action by legislatures. Meanwhile, the state legislators approving these gerrymandered maps for congressional seats often also control the redistricting process for their own legislative bodies, enabling them to remain entrenched.

The soft power of Donald Trump

The extra factor that Trump brought to the party was his extensive use of new media. Using mainly Twitter, supplemented by other outlets, he delivered his message directly to supporters without the intervention of political party or journalists and with an unceasingly aggressive and negative tone. He continued in a similar mode through his term in office. Trump complemented his use of alternative media streams with relentless attacks on traditionally trusted media sources. When challenged by CBS correspondent Lesley Stahl about his repeated attacks, Trump responded, "I do it to discredit you and demean you all so that when you write negative stories about me no one will believe you" (Mangan web). Fox TV provided a safe space complementing the Twitter feed and enabling Trump and his supporters to distribute their "alternative facts."

If any channels became concerned about the extremity and falsehoods in the administration's claims, cable and internet distribution provided alternatives. Newsmax, OAN, Parler, Telegram, Signal and other platforms created a mutually supportive "echo chamber" of media, where conspiracy theories flourish. While Democrats continue, in general, to rely on traditional media sources, Republicans are more likely to look elsewhere. Neither trusts the others' sources. Trump proved that a president who in one administration tells over 30,000 lies can, with the help of new media, shake the foundations of established party politics.

Weaponizing defeat: From big lie to big steal

The leadership of a political party might be expected to use any legitimate method to help it gain political power. It may be embarrassing to win an election without gaining the support of a plurality of voters, but the USA is not the only country where this can happen. If the victory comes within the pre-existing electoral regulations, it might ordinarily stimulate an examination of those regulations without the attempted insurrection in Washington DC on January 6th that might now be the direction of political debate in the USA. But the GOP's swerve away from the Republican National Committee's own 2013 *Growth and Opportunity Project Report* suggests otherwise. The party had already opted to leverage any potential bias in the system that might bolster its minority-based authority in the nation.

One speaker at the January 6th rally, John Eastman, then working with President Trump's legal team, had days before presented a six-point plan advising Vice-President Mike Pence to overturn the election while tabulating the Electoral College vote. Anticipating "howls" from the Democrats, Eastman pushed the idea that the decision could then be handed to the House of Representatives, where the 26 states controlled by Republicans could deliver the presidency to Donald Trump. Any thoughts that Trump's defeat and removal from office would swing the system back to a quiet equilibrium diminished quickly in the early part of President Biden's administration. Even in mid-2023, as America began to gear up for the 2024 presidential election, an article in *Forbes* reported that Trump loyalists remain unmoved, with 63 per cent of Republican-leaners in the electorate retaining the belief that President Biden did not win the election legitimately (Durkee web).

Challenges to the 2020 election results have repeatedly failed, having provided no evidence. One judge responded that "This claim, like Frankenstein's Monster, has been haphazardly stitched together" (Feuer and Montague web). A Republican financed audit of the ballots in Maricopa County, Arizona found no fraud, reporting that, if anything, Biden may have suffered from a small undercount. Audits nevertheless continued to be pursued by disappointed Republicans in Wisconsin and Pennsylvania, all proving fruitless from the complainants' perspective.

Former President Trump contrived to insist to his followers that the Arizona audit confirmed his belief that the election was rigged and demanded that Texas (a state that he won) initiate audits. Texas promptly did so in four counties. This constant challenging of the legitimacy of the election process echoes Trump's words to Lesley Stahl: "I do it to discredit you and demean you all so that … no one will believe you." (Mangan web) It sets the stage for continued attacks by Trump and his allies on future elections.

The Eastman memo, the legal challenges, the attack on the Capitol and the audits failed to overturn the legitimate election result. It did increase fears that a Republican Party that failed to deny the Big Lie, failed to distance itself from its insurrectionist supporters, continued to peddle conspiracies and to undermine science, attempted to undermine the elections process and used all the tools it could to maintain authority with the support of a minority of the nation who was threatening the long-term health of American democracy.

Voter suppression and Gerrymandering

Claims that the Democrats stole the 2020 presidential election also act to justify proposals that will reduce access to the ballot. The Brennan Center reported a blitz of over 400 bills introduced in 49 states with provisions that restrict voter access. This wave of restrictive proposals is, according to the Center, "in large part motivated by false and often racist allegations about voter fraud" (Johnson 2022: 195). Big Lie and Big Steal rhetoric feeds the suspicions that justify these voter suppression activities. The suppression acts largely against likely Democratic voting communities.

With the publication of the 2020 Census results many states moved quickly towards redistricting their constituency maps. Allen West, chairman of the Texas GOP, prevailed on his party members in that state's legislature to "realise this strategic opportunity and not concern themselves with 'fairness'" (Flahive web). The initial proposal from the Republican dominated Texas state senate "aims to insulate GOP incumbents from the state's rapidly diversifying electorate, while diminishing the voting power of Latinos and Blacks in the state" (Itkowitz *et al*). In the current combative atmosphere other states are followed suit, including some of those controlled by Democrats, further undermining the principle that all voters have approximately equal representation through the ballot box. The new boundaries once set remain part of the political landscape until 2032.

Culture wars and political polarisation

Texas also serves as case study regarding emerging and re-emerging "culture war" issues.

- Reproductive rights remain highly contested. The Guttmacher Institute reported that 2021 saw state legislatures introduce "a record number of abortion restrictions" (Nash 2022) including attempts to ban abortions after 6 weeks of pregnancy, and, in Texas, offering financial incentives for private individuals to take legal action to prevent abortions. The Supreme Court refused to delay implementation of the Texas law.
- Texas relaxed gun laws such that no licence, background check or training is needed in order to carry a handgun.
- In Colleyville, Texas the first black person to be made headteacher of a local school was suspended in an argument over teaching critical race theory.

Republican legislators in other states are attempting to ban CRT from schools and colleges.
- In September 2021 Texas Governor, Greg Abbott signed a law imposing new restrictions on voting hours, absentee voting and assisting voters.

These are "hot button" issues for many conservative voters. One Texas academic, Professor Albert Kauffman of St. Mary's University School of Law opined, "I think a lot of [Republicans] look at the demographics and say, 'We might not have more than four or six or eight more years in control so we probably ought to put in as many mechanisms as we can to protect ourselves long-term.'" (Mayes *et al* web).

The internationally important debate over public health responses to the Covid-19 pandemic has been hijacked in many states into arguments over ill-defined concepts of "freedom" or "liberty." Republican Representative Madison Cawthorn (NC) went so far as to warn that President Biden's mobile vaccine teams may be used to confiscate household bibles and guns, while Florida's Governor DeSantis promoted "Don't Fauci My Florida" merchandise.[8]

Senior Republicans in Congress frequently call on the Biden administration for bipartisanship without showing any inclination to enable it. One issue echoes back to the problems cited ten years earlier by David Moss. Senate Minority Leader Mitch McConnell vowed that "Republicans will not offer any more assistance to raise the debt ceiling," once again using the risk of government default and national (and international) fiscal chaos as a party-political grandstanding.

Dodging the demographics

But dependence on political tricks that allow a minority party to retain political power cannot be a long-term strategy. The latest census re-emphasises that political demographics appear more promising for the Democratic party's future. However, as investment guides always say, "past performance is not a guarantee of future results" and if anyone in the GOP is thinking in the long term, they may well be counting on slowing the impact of demographic change to give their party the opportunity to counteract the political shifts such change may bring, they may be wrong.

The midterm elections of 2022 offered the Republicans an excellent opportunity to take control of the US House and Senate. Only one seat needed to shift in

---

8   Gov. Ron DeSantis sells "Don't Fauci My Florida merch" New York Post, July 15, 2021 (nypost.com)

the Senate and the House election was fought using a constituency map expected to favour the Republican Party. In the event there was no great shift to the GOP. The June 2022 Supreme Court decision on abortion (*Dobbs v. Women's Health Organization*) and the rapid reaction anti-abortion forces in some states, may have stirred some voters against the conservative swing of recent elections. The Democrats clung on to the Senate. The Republicans took the House by a tiny margin. In both cases the party leaders, with their very small majorities of seats, risk being held hostage by individual members and small groups who are willing to exploit these tiny margins in order to exercise political leverage.

The Democrats' performance in the 2022 midterms also strengthened President Biden's position as his party's leading candidate for the 2024 nomination, in spite of his advanced age and his relatively low public opinion poll ratings. In the 2024 elections 23 of the 33 Senate seats up for election are held by Democrats and their allies, providing a huge target for Republican gains. With former President Trump a potential candidate for the Republican nomination, facing potential challenges from other Republican conservatives such as Senator Ron DeSantis (Florida), Senator Ted Cruz (Texas) and Governor George Abbott (Texas) this is likely to be another brutal, expensive contest.

In a warning that demographics may not come strongly to the aid of the Democratic Party in the long run, there were noticeable swings towards the Republicans in 2020 among some Latinos and other traditional Democratic voting groups – especially among males and voters without college education. Elaine Kamarck of Brookings commented that "Hispanics could turn out to be the Italians of the 21st century—family-oriented, hardworking, culturally conservative. If they follow the normal intergenerational immigrant trajectory rather than the distinctive African American path, the multi-ethnic coalition on which Democrats are depending for their party's future could lose an essential component" (Galston web).

In short, demographic and associated political change may benefit the Democratic Party, but if the Republicans can use their current systemic advantages to maintain authority for another decade, they may see the possibility of adapting and forging a new, stable, long-term conservative electoral coalition that does not need legerdemain to remain in office.

Conclusion: A "Path to re-equilibration," or nightmares to come?

The US-based organisation Freedom House noted a decline in the United States' *Freedom in the World* score of 11 points, from 94 to 83, over the past decade

(Repucci and Slipowitz web). Among the factors driving this change were political corruption and conflicts of interest, lack of transparency in government and punitive immigration and asylum policies. The Economist Intelligence Unit listed the USA in the category "flawed democracies" in its 2021 report, citing "extremely low levels of trust in institutions and political parties, deep dysfunction in the functioning of government, increasing threats to freedom of expression, and a degree of societal polarisation that makes consensus almost impossible to achieve."[9] (EIU 6)

The University of Virginia's respected Center for Politics released survey results in September 2021 indicating that 52 per cent of Trump voters and 41 per cent of Biden voters somewhat or strongly agreed that "the situation in America is such that I would favor [Blue/Red] states seceding from the union to form their own separate country."[10] The survey found common ground on several issues, but deep mistrust extending to the belief that elected officials from the other party "represent a clear and present danger to American democracy," a division that Larry Sabato called "deep, wide, and dangerous." ("Sabato's Crystal Ball") And that cannot be fixed by party leaders who rush to confrontation and manipulation, rather than co-operation and transparency.

A suite of essays in *PS: Perspectives on Politics* (June 2020) addressed these matters. Most were quite positive. Francis Lee argues that the USA offers more opportunity for populist candidacies than for populist political parties, but concludes that the US constitutional structure provides barriers to populist leaders undermining the structure of democracy. Kurt Weyland hypothesised that populist attacks on the system could prompt counter mobilisation that would revitalise US democracy. Michael Bernhard and Daniel O'Neill wrote hopefully of the "path to re-equilibration."[11] Stephen Skowronek and Karen Orren opined that contemporary pressures on American democracy may test its traditional resilience. "[T]his polity is awash in its principles" but is lacking "a reliable structure of decision-making authority."[12] Accordingly, American democracy may not be absolutely on the edge but it is at risk. It remains difficult to deny the possibility that William Riker's nightmares may yet be justified.

---

9   Democracy Index 2020 – Economist Intelligence Unit (eiu.com)
10 New Initiative Explores Deep, Persistent Divides Between Biden and Trump Voters – Sabato's Crystal Ball (centerforpolitics.org)
11 Whither America? | Perspectives on Politics | Cambridge Core
12 The Adaptability Paradox: Constitutional Resilience and Principles of Good Government in Twenty-First-Century America | Perspectives on Politics | Cambridge Core

References

American Political Science Association Committee on Political Parties. "Toward a More Responsible Two-Party System." *The American Political Science Review*. Vol. 44, no. 3, part 2:6. 1950. Web.

Bernhard, Michael and Daniel O'Neill. "Whither America?" *Perspectives on Politics*. Vol. 18, no. 2. 2020. Web.

Bronner, Laura and Nathaniel Rakich. "Advantage, GOP: Why Democrats have to win large majorities in order to govern while Republicans don't need majorities at all." *FiveThirtyEight*. 29 April 2021. Web.

Burnham, Walter Dean. *Critical Elections Revisited*. Lecture delivered at the University of Texas. 5 April 2006. Web.

Byler, David. "Republicans Now Enjoy Unmatched Power in the States. It Was a 40-year Effort." *The Washington Post*. 18 February 2021. Web.

Carnes, Nicholas and Noam Lupu. "The White Working Class and the 2016 Election." *Perspectives on Politics*. Vol. 19, no. 1, pp. 1–18. 2021. Web.

Committee for Party Renewal. "Policy Statement on the Role of State Parties." *PS: Political Science and Politics*. Vol. 21, no. 3, pp. 642–643. 1988. Web.

Dionne, E.J., Thomas Mann and Norman Ornstein. *One Nation After Trump*. New York: St Martin's Press. 2017. Print.

Durkee, Alison. "Republicans Increasingly Realize There's No Evidence of Election Fraud – But Most Still Think That 2020 Election Was Stolen Anyway, Poll Finds." *Forbes*. 14 March 2023. Web.

Economist Intelligence Unit. "Democracy Index 2020: In Sickness and in Health?" *Economist Intelligence Unit*. 2021. Web.

Feuer, Alan and Zach Montague, "Over 30 Trump Campaign Lawsuits Have Failed. Some Rulings Are Scathing." *The New York Times*. 25 November 2020. Web.

Flahive, Paul. "Texas GOP Chairman Calls for Increased Gerrymandering." *Texas Public Radio*. 11 January 2021. Web.

Galston, William. "New 2020 Voter Data: How Biden Won, How Trump Kept the Race Close, and What It Tells Us About the Future." *Brookings*. 6 July 2021. Web.

Green, Joshua. "The Rove Presidency". *The Atlantic*. September 2007. Web.

Hiatt, Fred. "An Option for Obama: Leadership." *The Washington Post*. 31 October 2010. Web.

Itkowitz, Colby, Harry Stevens and Adrian Blanco. "Texas GOP lawmakers' Redistricting Map Protects Congressional Incumbents While Avoiding a New Latino-majority Seat." *The Washington Post*. 27 September 2021. Web.

Johnson, Dennis W. *Campaigns, Elections and the Treat to Democracy. What Everybody Needs to Know*. New York: Oxford University Press. 2022. Print.

Lee, Frances E. "Populism and the American Party System: Opportunities and Constraints." *Perspectives on Politics*. Vol. 18, no. 2, pp. 2020. Web.

Luce, Edward. "'One Nation After Trump' by Dionne, Ornstein and Mann." *The Financial Times*. 9 October 2017. Web.

Mangan, Dan. "President Trump Told Lesley Stahl He Bashes Press 'to demean you and discredit you so … no one will believe' Negative Stories About Him." *CNBC*. 22 May 2018. Web/Youtube.

Mann, Thomas and Norman Ornstein. *It's Even Worse Than It Was*. New York: Basic Books. 2016. Print.

Mann, Thomas and Norman Ornstein. *The Broken Branch: How the Congress is Failing America and How to Get It Back on Track*. Oxford: Oxford University Press. 2006. Print.

Mann, Thomas and Norman Ornstein. "The Broken Branch: How the Congress is Failing America and How to Get It Back on Track." *Brookings*. 27 June 2006. Web.

McCain, John. "Concession Speech." *NPR*. 5 November 2008. Web.

Mayes, Brittany Renee, Lesley Shapiro and Zach Levitt. "Why Texas's laws are moving right while its population shifts left." *The Washington Post*. 20 September 2021. Web.

Moss, David A. "Fixing What's Wrong with US Politics." *Harvard Business Review*. March 2012. Web.

Nash, Elizabeth. "State Policy Trends 2021: The Worst Year for Abortion Rights in Almost Half a Century." Guttmacher Institute. Updated 5 January 2022. Web.

REDMAP (The Redistricting Majority Project). Republican National Committee's 'Growth and Opportunity Project' Report – DocumentCloud. http://www.redistrictingmajorityproject.com/. Web.

Republican National Committee. *Growth and Opportunity Project Report*. 2013. Web.

Repucci, Sarah and Amy Slipowitz. "Democracy Under Siege." *Freedom House*. 2021. Web.

Riker, William. *Liberalism Against Populism*. Long Grove, IL: Waveland Press. 1982. Print.

Roy, Avik. "No, Trump Didn't Win 'The Largest Share of Non-white Voters of any Republican in 60 Years.'" *Forbes*. 9 November 2021. Web.

Skowronek, Stephen and Karen Orren. "The Adaptability Paradox: Constitutional Resilience and Principles of Good Government in Twenty-First-Century America." *Perspectives on Politics*. Vol. 18, no. 2, pp. 354–369. 2020. Web.

Uslaner, Eric. *The Decline of Comity in Congress*. Ann Arbor: University of Michigan Press.1993. Print.

UVA Center for Politics. "New Initiative Explores Deep, Persistent Divides Between Biden and Trump Voters." *Sabato's Crystal Ball*. 30 September 2021. Web.

Weyland, Kurt. "Populism's Threat to Democracy: Comparative Lessons for the United States." *Perspectives on Politics*. Vol. 18, no. 2, pp. 389–406. 2020. Web.

Other internet sources:

Critical Realignments Revisited (utexas.edu)

The Broken Branch: How Congress Is Failing America and How to Get It Back on Track (brookings.edu)

https://hbr.org/2012/03/fixing-whats-wrong-with-us-politics

Election Exit Polls 2020 – *The New York Times* (nytimes.com)

Transcript Of John McCain's Concession Speech: NPR

Advantage, GOP | FiveThirtyEight

The RSLC Redistricting Majority Project – REDMAP

Gov. Ron DeSantis sells "Don't Fauci My Florida merch" *New York Post*, July 15, 2021 (nypost.com)

Democracy Index 2020 – Economist Intelligence Unit (eiu.com)

New Initiative Explores Deep, Persistent Divides Between Biden and Trump Voters – Sabato's Crystal Ball (centerforpolitics.org)

Whither America? | Perspectives on Politics | Cambridge Core

The Adaptability Paradox: Constitutional Resilience and Principles of Good Government in Twenty-First-Century America | Perspectives on Politics | Cambridge Core

Eduard Vlad

The Great Reset and Its Pandemic Narratives

Abstract: The current essay starts from conceptualizations and events that became visible in what might be called one day "the Covid-19 age," or even the age of "the pandemic narratives of the Great Reset." This essay places Schwab and Malleret's controversial *Covid-19: The Great Reset* in relation to the various contexts in which it appeared for a better understanding of the outrage that it occasioned, also giving rise to a host of conspiracy theories as the raw materials and the pins and bolts of what are called here "pandemic narratives" in their close relation to the already mentioned conspiracies.

Introduction: Conspiracy theories and pandemic narratives

It is useful to see whether the apparently "peaceful coexistence" of conspiracy theories and pandemic narratives has any relevance to the examination of significant aspects of the age of the Great Reset, which the title of this essay has already resuscitated at a time when *I Am Ukrainian* tends to remind one of President J.F. Kennedy's famous statement, *Ich bin ein Berliner*, and of its geopolitical significance. By coincidence, in the postwar age, the most influential conspiracy theory also had to do with JFK's assassination. Conspiracy theories have been around ever since, as well as pandemic narratives, if one also considers the worldwide "containment" of Communism that became a geopolitical imperative in the early years of the Cold War. The "red plague" having to do with the infectious and contaminating nature of Communism is related to such figurative language and such pandemic narrative discourse. In such cases as this Cold War containment narrative, the concept of conspiracy theory does not appear to be of much use. The militancy of Communism had been obvious and very public ever since Marx's *Communist Manifesto*, while conspiracy theories involve a great deal of secrecy. Contamination, containment, and resistance to it, as well as conspiratorial elite circles will nevertheless feature prominently in the Covid-19 age, which explains why pandemic narratives and conspiracy theories help each other invade the many global pathways of information, misinformation, disinformation.

    Did the medical world and the public know much about the thousands of variants of the new Covid-19 in the summer of 2020? Obviously, no. The new viral pathogen and the pandemic were new realities urging the scientific world to move faster than during previous similar outbreaks. The classification, research and development of new therapies is a central preoccupation of such institutions as the WHO. Similar preoccupations have always been subject to contestation and heated debates, such as the classification of the pandemic alert phases in the wake of the 2009 H1N1 flu epidemic (Abeysinghe 905). There has always been suspicion of such classificatory and explanatory work on new versions of infectious diseases and of possible remedies, especially fueled by a long tradition of vaccination hesitancy discourse in the public space, lately strongly amplified by the social media. What is more, this is the fertile ground for conspiracy theories about dishonest scientists, rapacious and unscrupulous pharmaceutical and vaccine-producing companies, occult business people and corporations deriving huge profits from the confrontation with an invisible and hardly known biological enemy.

If one cannot clearly identify the biological enemy and the arch-villains of the global elites conspiring against the rest of humanity, at least one can safely say that definitions of conspiracy theories are more easily to be arrived at. Conspiracy theories are aptly seen in just one sentence (albeit a very long one, here abridged) by Jovan Byford as

> [...]large scale, dramatic social and political events (such as the AIDS epidemic, the assassination of John F. Kennedy or 9/11); for explanations that do not just describe or explain an alleged conspiracy, but also uncover it and in doing so expose some remarkable and hitherto unknown "truth" about the world (such as that the Illuminati orchestrated the French Revolution or that the Bush administration had a hand in 9/11); and for accounts that allege the existence of a plot with nefarious and threatening aims (to destroy Christianity, establish the New World Order ...) (Byford 21).

Pandemic narratives as part of a global, largely contradictory pandemic discourse, derive some of their power and urgency from conspiracy theories. The phrase "pandemic narratives" refers here to both the narratives about the Covid-19 global visitation and to the multi-faceted viral shape, scope and power of penetration of discourse in the public space today, particularly with narratives dramatizing the pros and cons of the eco imperatives of the current times. These narratives feature voices as diverse as those of Henry A. Giroux and Joseph Mercola, Klaus Schwab, Prince Charles (not Charles III at the time the current essay was written] or Pope Francis. The most comprehensive narratives include and revolve around public health, economic, political, and geopolitical concerns.

The shape and content of the pandemic narratives were further complicated by such important moments as the 2020 Presidential elections in the U.S., which added extra divisiveness in terms of specific political agendas. These agendas highlighted prevailing camps and ideologies such as *America First*, a 21st century echo of a dubious pre-World War II movement, and *Black Lives Matter*, a revival, in the context of George Floyd's murder, of the current imperatives of the previous Civil Rights Movement of the 1950s and 1960s. To make matters worse, the last months of 2020 introduced QAnon and its story about Trump's providential power to thwart the horrible secret plot of the Great Reset, if the Republican candidate is re-elected, leading to the storming of the Capitol on January 6, 2021.

These pandemic narratives owe their shape and scope to conceptualizations of what Jan van Dijk and Manuel Castells call the network society. In *The Rise of the Network Society*, his 2010 reworking of the now classic *The Network Society*, Manuel Castells examines the ways in which the information age has transformed economies, societies, cultures, featuring new patterns of doing business and organizing things, but also creating equally new challenges because of the increasing integration and interconnection of the global economic framework. Castells metaphorically claims that his theoretical project "swims against streams of deconstruction, and takes exception to various forms of intellectual nihilism" (Castells 4), the author firmly believing in rationality.

Yet networks and webs are not necessarily obeying reason. Any world wide web "irrationally" attracts viruses, acting very much like interconnected global communities spreading new viruses and conspiracy theories alike. If conspiracy theories are "paradigmatically irrational" (Coady 1), so are pandemic narratives. They incorporate in their complex, sometimes contradictory patterns, aspects of one of the defining components of what Douglas Murray (2019) calls "the madness of crowds." The phrase, although intended by Murray to refer to irrational ways of dealing with aspects of contemporary group identity, might aptly describe the fascination with, and appeal to, unknown dangers that conspiracy theories feature. Conspiracy theories linked to the Covid-19 pandemic narratives had long been part of popular culture. Among them, here are some acknowledged in Jan-Willem van Prooijen's *The Psychology of Conspiracies*, published as early as 2018, apparently "prophesying" impending theories about vaccines and climate change that would soon exert a pandemic effect, comparable to that of Covid-19 itself:

> Climate change is a hoax perpetrated by the Chinese. The pharmaceutical industry hides evidence that vaccines cause autism. [...]. These are conspiracy theories that were propagated during the 2016 presidential campaign of Donald Trump. Scientists, journalists, policy makers, and other critics portray these theories as naïve, far-fetched, not supported by evidence, or simply ridiculous. But why then do so many people believe such conspiracy theories? (van Prooijen, vii)

If sometimes irrational political fanaticism might explain the appeal of certain conspiracy theories, pandemic narratives appear to be able to incorporate various narrative strands appealing to a variety of audiences and serving the needs and interests of various agencies. Thus, at certain moments and in certain places and spaces during the Covid-19 age, Trump supporters and detractors, anti-environmentalists and environmentalists alike, liberals and conservatives, might feel tempted to give credence to the existence of a sinister conspiracy hiding behind the façade of the Great Reset initiative.

Information and technological networks on the one hand and cultural webs on the other produce viruses of some kind or other, one major characteristic that they share. Since a bewildering array of things become viral these days, pandemic-like, rather than network-like, a pandemic narrative is a suitable visualization of discourse displaying extreme connectivity and contagion, but also involving "contradictory pluridirectionality," as it were, often wreaking havoc largely by means of the sensationalist messages disseminated by influential voices in the social media.

The evocative power of the noun *pandemic* used as an adjective is thoroughly exploited by Henry A. Giroux in his 2021 volume, *Race, Politics and Pandemic Pedagogy: Education in a Time of Crisis*. As the author clarifies it from the beginning, the time of crisis he evokes and examines starts on the day in late January 2020 when the World Health Organization officially declared the new virus a Public Health Emergency of International Concern, while the official declaration of the pandemic came on March 11th. This critical time coincides with what is referred to in the current essay by the phrase "the age of the Great Reset." Giroux's book goes on to chronicle subsequent developments up to the failure of Donald Trump to secure a second presidential term.

Giroux notes with alarm the rising levels of state repression worldwide, but focusing on the US, with the occupation of American cities by military troops and similar developments, as well as the global backlash to such negative developments. Giroux mirrors views of important American Critical Race theorists, such as Derrick Bell and Robin DiAngelo on systemic and institutional racism exacerbated by the murder of George Floyd in May 2020. He notes that the Covid-19 pandemic and its ensuing crisis heralding a massive economic collapse have unveiled dramatic dimensions of poverty and its attending suffering that had previously been somewhat obscured by the triumphant discourse of liberal capitalism. This Covid-19 age, Giroux claims, gave rise to political formations that changed the kind of debate concerning the challenge of state violence and of the emergence of fascist politics during the Trump administration, also unmasking the mechanisms of a failed state which, under the aegis of neoliberalism and of its pandemic pedagogy, piles up evil upon evil. More, this pandemic pedagogy "reads the world largely through racialized categories, the purification of national cultures, and the language of privatization and commodity consumption" (Giroux 4).

Giroux goes on to introduce his view on this so-called pandemic pedagogy announced by his book's title. He speaks about the obnoxious nature of the ideological virus-plague that stemmed from a politics of depoliticization linked to a series of market-based assumptions and pedagogical practices heralding the already mentioned pandemic pedagogy. This form of pandemic pedagogy attacks and subverts key ideas, values, desires, and assertions of agency empowering people to become actively involved citizens. This pandemic pedagogy is meant to articulate patterns of agency and identification which prompt individuals to give credit to and participate in a system which further concentrates power in the hands of those whom sociologist C. Wright Mills had called "the power elite," stressing the establishment of an economic order based on "a private corporation economy" (Mills 275–276).

Giroux introduces in the pandemic narratives under examination in what is to follow the almost conspiratorial presence of the elites that, ambivalently, *both* endanger the health of the employees by urging them to work during the highly contagious stages of the pandemic *and* contain them, imposing lockdown and curtailing their basic liberties. Either way, whether they endanger the people's health or they deprive them of their constitutional rights, the same sinister elites, hiding beyond or above the official governments, appear to bear the blame for the acknowledged vacillations in the dilemma *the economy first* vs *public health first* that various parties and prominent leaders engaged in, with the notorious cases of Boris Johnson and Donald Trump switching between opposite positions.

Pandemic narratives assuming a conspiratorial appearance maximize the scope of the catastrophic influence that the malevolent elites had in the developments of the Covid-19 age of the Great Reset. Joseph Mercola and Ronnie Cummings, in a volume that, judging by the title, appears to be fully enlightening, *The Truth about Covid-19: Exposing the Great Reset, Lockdowns, Vaccine Passports, and the New Normal*, also indict the beneficiaries and main conspirators: "Covid-19 has empowered the global elite more than ever before to manufacture lies and untruths." Among the mentioned liars and speculators, in the opinion of the two authors, feature prominently Facebook, Google, Microsoft, Amazon, the World Health Organization, and, last but not least, Bill Gates himself. Their major crimes are "fearmongering, political polarization, social engineering – all wrapped in a disguise of protection" (Mercola & Cummings 1). The two authors call for a global awakening in what appears to be "the call of our lives." Is a "global awakening" and the already invoked "great reset" mortal enemies? A pandemic narrative amounting to a battle cry against the Great Reset occupied most of the time already mentioned by Giroux, but also extended well into 2022, although with decreased intensity. However, there is a possibility that, not for the first time in history, one takes up arms against a feared enemy that one barely knows and often misrepresents, usually in typically tabloid journalistic style as revealed from the very beginning by the apocalyptic title of Mercola and Cummings's 2021 volume.

Pandemic settings, entities, groups

Arguably, the most extravagant thread in the complex, sometimes contradictory pandemic discourse is that the Great Reset is a secret, conspiratorial plan hatched by the global elites. These elites have invariably the same interests, they do not engage in fierce capitalist competition. They somehow created and then managed the Covid-19 pandemic to make people poorer and more dependent on them, to curtail their freedom, to establish totalitarian states enforcing universal Orwellian surveillance.

Among the most fearsome enemies featuring in the conspiratorial, pandemic narratives of the Great Reset is undoubtedly the World Economic Forum, although, to do Giroux justice, he does not include it among the evildoers of the age on which his book, and the current essay, focus. In times of global crises, fears of worldwide plans involving mysterious and influential transnational agents are bound to strike many a sensitive chord with numerous intelligent individuals who, lacking thorough and extensive research and exploration of multiples sources and perspectives, decide that they are not gullible enough to accept official narratives coming from the political or the scientific establishments. The World Economic Forum has attracted fears having to do with a set of interwoven global concerns. One of these is the overall ecological concern linked to the consequences of the increasing violence that humankind is perpetrating against the environment. The main actors are to be identified and challenged, and anything and anybody in the shadow of such impressive labels as the World Economic Forum, the Global Redesign Initiative, or, worst of all, the Great Reset, to all intents and purposes, are to be exposed. The fact that WEF and the Great Reset advocate the switch to greener sources of energy is not usually considered, and the reasons why might require more research (and more essays) on the influence of oil and gas providers on pandemic narratives and on the terrible conspiracies of the "green" elites.

Concerns about the threat posed by human societies to the environment these days are based on fact. They are legitimate, and developing an eco-consciousness is an imperative, as the theme of the conference that prompted this text states. Apparently, everyone is paying lip service to this imperative. However, a multiplicity of competing interests and priorities divide each society and each nation, the whole world. An increasing degree of interconnection and communication, prominent figures, ideas, slogans are just as necessary for the competing parties as the eco imperatives themselves. If one wants to engage in this complex maze of pandemic narratives having to do with the effects of pollution on climate change, with the origins and consequences of Covid-19, with plans to control the world by means of both fears and concerns about the pandemic and with plans for a mysterious Great Reset, one must start somewhere.

Elusive, comprehensive mazes, networks have to be narrativized somehow, they have to be turned into a clear story starting somewhere with one or several central characters, with possible motivations and consequences. If one wants to challenge the sinister threat posed by the Great Reset, the best way is to choose specific characters and to associate them with some apparently unacceptable ideology. The easiest way is to choose Klaus Schwab, both the founder of the World Economic Forum and the author of the 2020 volume, *Covid-19: The Great Reset*. How about attributing to this central character in this pandemic narrative such an apparently shocking slogan as *You'll Own Nothing and Be Happy*? It is a shocking slogan if taken literally: nothing, nothing at all? Answering this final question is probably the easiest response in this vast pandemic maze, and it will be postponed until the discussion of this slogan is put in a more comprehensive context later on in this essay. For the time being, a discussion of the central character of the narrative, of his entourage and of his central text are worth undertaking.

The central character and his "conspiratorial" company

First, let's imagine the way some of the challenges of the Great Reset appear. Let's say something about the QAnon people that made such a big fuss around the 2020 Presidential elections in their desperate support of Donald Trump against the globalist elites. Let's imagine how individuals like the above-mentioned see Klaus Schwab, the figurehead of the World Economic Forum, are likely to be perceived by many people in the public space. The fact that he holds a doctorate in Economics from the Swiss Federal Institute of Technology and an MA from the Kennedy School of Government at Harvard University might be of less interest to the QAnon people. Schwab is unglamorously advanced in years, over

80, and very German-looking. He speaks English with a thick Teutonic accent; in addition, the various meanings of the word *schwab* in "dirty" American slang are not at all complimentary. It adds another touch to the xenophobic prejudice against the German-Swiss author. What is more, one might associate his image with the images of the "enemy": Joe Biden, George Soros, Prince Charles (future Charles III), Pope Francis, all of them viewed as "fascists" advanced in years, plotting against honest, ordinary people everywhere, such as the above-mentioned QAnon militants.

One such association of "enemies" can be seen in videos like The Great Reset: Joe Biden and the Rise of 21st-Century Fascism – YouTube. So, it is Schwab, Biden, Soros and 21st century Fascism rolled into one, what can be scarier and more detestable for such respectable people as the QAnon militants? One question is worth pondering, though: which regimes openly declare such figures as the equally unglamorous George Soros as the enemy of their peoples and their cultures? The Open Society project promoted by Soros in an attempt to challenge totalitarian thought was not met with much enthusiasm in a number of otherwise very democratic states like the Russian Federation and Hungary, an understatement which does not need any further evidence for anyone living in 2023.

Is Klaus Schwab the Most Dangerous Man in the World? – YouTube is another video asking a rhetorical question, in the opinion of its author, Trump militant and sympathizer extreme-right winger commonly known as *awakenwithjp* (the real name is J.P. Sears, comedian and businessman, whose job is to satirize the American left). If one is fully awake, the answer is obviously YES. Again, Schwab is here associated with President Joe Biden, with other liberal figures linked to the "evil elites." A telling illustration involves Leonardo DiCaprio sitting close to President Biden during a public event in the same video. Significantly, DiCaprio has been distinctly left leaning and environmentally minded, having set up an environmental foundation, having gone so far as to get Pope Francis involved in the making of an environmental film. Daniel Bodansky confirms DiCaprio's and Pope Francis's common environmental pursuits, commenting on the papal encyclical, *Laudatio Si*, made public in May 2015.

To add to the "sinister group" around Schwab, let's evoke another hostile right-wing video to Schwab and the Great Reset: Prince Charles is now selling his 'eco-fascist fantasy of the Great Reset' – Bing video. Apart from Klaus Schwab advertising the Great Reset in thickly accented English, the video shows Prince Charles speaking in very good German about the same "eco-fascist" project.

In addition to one particular central character, who might be either Klaus Schwab or Prince Charles (King Charles III), but George Soros or President Joe

Biden would equally do, an important anti-Great Reset narrative also needs settings, even in an age in which almost everything is deterritorialized. One could choose to link Schwab and his World Economic Forum to a place that fans of Gothic narratives might like. More precisely, Klaus Schwab founded the World Economic Forum half a century ago (1971) in a small mountain village in the canton of Geneva, Switzerland, by the name of Cologny. In a conspiracy theory one can easily make a connection between Schwab, the WEF, Frankenstein's monster and the picturesque village of Cologny. In 1816, a year afflicted by an even more catastrophic global phenomenon than the Covid-19 pandemic, Mary Shelley was inspired by the darkness that enveloped the sky for several months because of the massive volcanic ashes obscuring the sky in both Cologny, where the Shelleys, Byron and their company resided, and the whole world. She came up with her terrible and horrible story about science going utterly wrong. So, it is in Cologny that Frankenstein's monster came alive, very much like the "monstrous" World Economic Forum. It is true that the forum usually holds its annual meetings in the opposite, eastern corner of Switzerland, in Davos, a skiing resort which has acquired sinister, Gothic associations for such anti-globalization organizations as *Public Eye on Davos*.

The reassuring realization is that a *forum*, like the Roman Forum of antiquity, is an open area of discussion, not a secret conspiratorial den, in which important political leaders, businesspeople and civil society activists debate issues of global concern. What is more, "the public eye" is on Davos, all its proceedings and debates being available online.[13]

The ideological weight of the "Gothic" narrative of the Great Reset

So much for the Gothic settings, Cologny and Davos, linked to Schwab and his "monster," the World Economic Forum, as suitable places of origin for the Great Reset narrative. Critical readers today, apart from a good plot, would also try to uncover a narrative's ideological weight. It is convenient to locate it in the book promoted by the founder of the WEF on the occasion of the 2022 Davos meeting, the much talked about, arguably hardly ever read (by the "conspiracy theorists"), *Covid-19: The Great Reset*.

It is worth stressing that the book launch had coincided with Prince Charles launching the 2020 Great Reset initiative through his video

---

13 https://www.weforum.org

#TheGreatReset – YouTube. Those interested might check for themselves if the environmentally minded message read by the man who was then the heir to the British throne is actually selling the "eco-fascist fantasy" that one of the anti-Great Reset videos mentioned above announced. As a matter of fact, the message that Prince Charles wanted to promote is that each crisis, including the one caused by the pandemic, is a time which offers opportunities for reconsidering global economic, social and political coordinates, an opportunity for the creation of a fairer, more resilient and sustainable world in what may be called a great reset. The entirely new sustainable industries that Prince Charles dreamed of are based on the harnessing of solar and wind power, rather than on energy obtained from natural gas and oil. Does this look bad in the context of the current Russian invasion of Ukraine and of Europe's dependence on one particular oil and gas supplier? Prince Charles's message might sound very nice to most Europeans today – is there a terrible conspiracy, a hidden agenda lurking in the contemporary eco-fascist heart of darkness? It so happens that to the ideological weight of the Great Reset one might feel tempted to add one minor detail: the transition toward renewable sources of energy appears to be not only an ecological imperative, but also a geopolitical imperative for most European countries.

The ideological weight is fueled not only by the general eco and geopolitical imperatives of the age, but also by the expertise of the authors of *Covid-19: The Great Reset*. The book is co-authored with Tierry Malleret. Unlike Klaus Schwab, Malleret has a very French name, having studied at the Sorbonne and the Ecole des Hautes Etudes en Sciences Sociales, Paris, and at St. Antony's College, Oxford, with an MA in Economics and History and a PhD in Economics. Malleret was instrumental in the founding of the Global Risk Network two decades ago, meant to study the emergence of five sets of worldwide risks: economic, geopolitical, environmental, social, technological. In the book, Malleret and Schwab aim at examining not only these risks, but also the opportunities arising in the wake of the massive destruction worked by the pandemic and its attending economic crisis. Does that look very fascist to commonsensical people?

Apparently, Schwab and Malleret's Great Reset manifesto is far from heralding or facilitating the emergence of a fascist, totalitarian, socialist world government as most critics of the global elites fear. The text acknowledges the indisputable fact that the global crisis caused by the Covid-19 pandemic brought monumental, unprecedented disruption, causing deep economic, social, environmental concerns, while threatening the health and welfare of a large section of the world's population. However, Schwab, Malleret and their Great Reset ideas have stirred an unprecedented series of violent responses from vaxxers and anti-vaxxers, conservatives and liberals, environmentalists and a host of other groups with various orientations, all amounting to a highly contradictory set of pandemic narrative sub-plots and threads. One is likely to be curious about the horrors that the book *Covid-19: The Great Reset* might hide within its covers.

Schwab and Malleret are fully aware that epidemics are bound to instill fear, anxiety, and moral panic, thus challenging social cohesion and the capacity of communities to cope with critical situations. They are divisive and traumatizing, as they compel people to fight an invisible enemy. Some are tempted to believe, rather than think, that the authorities are agents of oppression and totalitarianism when they impose measures of containment meant to keep large communities safe from infection. If the authorities are not imposing containment, then opposition parties will claim that public health is neglected to keep the economy going. The already mentioned dilemma (which, as it will be claimed in Schwab & Malleret, is "a false trade-off") is again invoked: which has priority, public health or the economy in the Covid-19 age?

Aware of all these serious problems, the authors first aim at sketching the conceptual framework necessary for the definition of the characteristics of the contemporary world, which are those of a globalized and globalizing world: interdependence, velocity, and complexity. If the interdependence and velocity of the globalized world are taken for granted by almost everyone today, the definition of complex systems is worth consideration:

> Complex systems are often characterized by an absence of visible causal links between their elements, which makes them virtually impossible to predict. Deep in ourselves, we sense that the more complex a system is, the greater the likelihood that something might go wrong and that an accident or an aberration might occur and propagate (Schwab and Malleret 32).

This largely explains high degrees of unpredictability, both the difficulty of grasping current developments and the contradictory, largely erratic conglomerate of stories interwoven in what the current essay calls "pandemic narratives." Like pandemics themselves, pandemic narratives are made up of various, often very disparate pieces of discourse, relating to public health, sociology, psychology, politics, geopolitics, civil rights, driven by as diverse agencies as governments, nongovernmental organizations, but also by influencers, bloggers, activists of all denominations engaged in a huge global debate. This is the main connection between the "specificity" of pandemic narratives and one of the three defining characteristics of the contemporary age, as mentioned by the authors.

In Schwab and Malleret's volume, the first dimension of the *macro reset* is the economic reset. The economic dimension of the Covid-19 age, so to speak, is infinitely faster, far more interconnected and complex now than at other equally dramatic times, such as those of previous catastrophic historical episodes, world pandemics, world wars. In addition to great devastations and loss of lives, all these catastrophes led to revolutionary changes in the economy and the social fabric of the various countries that they visited. Their destructive and "creative" consequences partially illustrate what Schumpeter describes in Chapter VII of his *Capitalism, Socialism and Democracy* as "creative destruction" (Schumpeter 81–86).

Challenges to the high degree of uncertainty surrounding the new virus with the fear and anxiety that influence economic behavior require new and bold scenarios linked to what the title of the project is all about: an economic reset as part of the Great Reset. This global economic reset should dismiss what the authors acknowledge as a widely shared perception: "the economic fallacy of sacrificing a few lives to save growth" (Schwab & Malleret 42). This is seen as a false trade-off, since loosening constraints and restrictions in order to facilitate normal economic activity is bound to lead to more people becoming ill and incapacitated. This would negatively affect the normal course of the economy while at the same time imposing a greater burden on the health system, among other negative consequences. It goes without saying that there is only one conclusion that the book promotes, despite the wild imaginings of conspiracy theorists focusing on it: "governments must do whatever it takes and spend whatever it costs in the interests of our health and our collective wealth for the economy to recover sustainability" (Schwab & Malleret 44). The return to growth might be slow, but this will prompt societies to pause and ponder what matters more, rapid growth or sustainable development involving a fairer, greener future. Another economic challenge contemplated in mid-2020 is the ability of the US dollar to preserve its favored hegemonic position, a question to be addressed in the new geopolitical uncertainties, and an answer to such a question from today's vantage point might be easier to supply, in connection with other attending questions, risks and opportunities that have emerged.

In the societal reset section of the book, the return of "big" government is another significant development, a mixed blessing in the context of the coronavirus pandemic that contributed to boosting the power of the authorities at the expense of the individuals. Some will see the advantages of a big and protective state, others will be afraid of their liberties being gradually curtailed by a state exerting ever increasing surveillance and control over its "subjected subjects."

Covid-19: The Great Reset may be seen with critical eyes by those who read too harshly some of the imperatives featuring in the third section, "Individual Reset." It is about redefining our humaneness, about the better, un-fallen angels in our nature, about major moral choices, as well as about stimulating creativity, controlling or not the revenge consumption after the pandemic, as well as the universally acknowledged ideals of all environmentally minded groups and individuals across the globe. Will there be a great reset at the individual level?

Since the book covers so much and the space here is limited, one final examination will move again from the individual to the geopolitical level. Although what the book sees is already out of date, it is a good starting point for subsequent discussions. What still stands is that the geopolitical reset, in the opinion of Schwab and Malleret, will have to address such thorny issues as globalization and the resurgence of nationalism and populism, aspects of global governance restraining the relative sovereignty of individual states, as well as, on center stage, the growing geopolitical competition between the US and China in the context of a "new type of Cold War." This new geopolitical rivalry dramatizes the current hegemon and its rising challenger, China. The latter, more insidiously than another competitor, Russia, does not appear to be willing to impose its ideology too openly and obtrusively, let alone engage in special military operations. At least for the time being.

The limitations of the 2020 Schwab and Malleret volume are inevitable in the dynamic age of the Great Reset, which has now been superseded by another, equally critical age. The two authors writing in July 2020 could not foresee the events following the February 2022 invasion of Ukraine, the attending new developments in the geopolitical arena. China will now think hard before openly siding with Russia as a possible supporter of its future global supremacy at the expense of the current hegemon. America will reconsider the importance of Europe as one of its main geopolitical pillars, while the EU may become more closely linked to American foreign policy aims and more willing to consider one of the eco imperatives of the Great Reset: the necessary transition from the "dirtier" forms of energy supplied by coal, natural gas and oil toward greener, more sustainable forms. This eco imperative of a cleaner, greener world will now be associated with the more down-to-earth imperative of putting an end to the EU's dependence on energy sources from Putin's Russia.

Conclusion

The focus of this essay, inevitably limited in scope, was on Schwab and Malleret's "outrageous" book, as part of a larger Great Reset initiative of debates initiated within the framework of a forum of discussions, not by a totalitarian entity. Around it, several pandemic narratives have taken shape, dealt with in the first part of the current text. From what little of the book has been discussed, it is hoped that at least a few fears harbored by people sensitive to conspiracy theories have been addressed. The most important one is about a risk during the pandemic, rather than a prerogative announced for the Great Reset, the possible boosting of the state's size and power and the possibility of increased control over the individual. Very few among the environmentally minded people who worry about climate change would have objected to the imperative of moving toward greener sources of energy.

The adjective "outrageous" above and the host of conspiracy theories imagined by influential social media figures as diverse as J.P. Sears and Russell Brand is largely due to a certain (mis) understanding of a famous Great Reset slogan. In all the videos mentioned in this essay and in many more "You will own nothing and you will be happy" is attributed to Klaus Schwab as part of his Great Reset plan. It is true that it became one of the "eight predictions for the world in 2030," but it comes from an essay written in 2016 by a very young environmentalist Danish politician, Ida Auken. If one takes this catchphrase out of the context of the powerful figures gracing the Davos event in 2020 and interprets it as Ida Auken originally meant it, one will realize its actual meaning. No one in 2030, it is hoped, will deprive anyone of their property, but a transition from buying and owning a large number of cumbersome things in one's house, garden, garage, etc., toward hiring almost everything at cheap prices and returning them to various service companies is a good option. This will avoid waste, while increasing comfort. A good interpretation of Ida Auken's slogan is, "you will have more money, because you will buy fewer things, while hiring more things." A good slogan is a striking one, sometimes featuring exaggeration. "You will own nothing," however, appears to have scared people who had far more than something. Attributed to Schwab and seen as part of a conspiracy of the elites to deprive most citizens of *all* their private property, the slogan was anathema to some conservatives and liberals alike in a capitalist world in which property is sacred. "If this be error…," one should refer to Ida Auken herself, thus checking, once again, how important contexts and original sources of information are.

References

Abeysinghe, Sudeepa. "When the Spread of Disease Becomes a Global Event: The Classification of Pandemics." *Social Studies of Science*. Vol. 43. No. 6 (December 2013): 905–926. Web.

Byford, Jovan. *Conspiracy Theories: A Critical Introduction*. Houndmills, Basingstoke, UK and New York: Palgrave Macmillan, 2011. Print.

Castells, Manuel. *The Rise of the Network Society*. Second edition. With a new preface. Malden, MA and Oxford: Blackwell, 2010. Print.

Coady, David. "An Introduction to the Philosophical Debate about Conspiracy Theories." *Conspiracy Theories: The Philosophical Debate*. Ed. David Coady. Aldershot, Hampshire, UK and Burlington, VT, USA: Ashgate, 2006: 1–12. Print.

Giroux, Henry A. *Race, Politics and Pandemic Pedagogy: Education in a Time of Crisis*. London/ New York/ Oxford/ New Delhi/Sydney, 2021. Print.

Mercola, Joseph and Ronnie Cummings. *The Truth about Covid-19: Exposing the Great Reset, Lockdowns, Vaccine Passports, and the New Normal*. Foreword by Robert F. Kennedy Jr. London: Chelsea Green Publishing, 2021. Print.

Mills, C. Wright. *The Power Elite*. New Edition. With a New Afterword by Alan Wolfe. Oxford and New York: Oxford University Press, 2000. Print.

Murray, Douglas. *The Madness of Crowds: Gender, Race and Identity*. London, Oxford, New York, New Delhi, Sydney: Bloomsbury Continuum, 2019. Print.

Padan, Carmin. "Coronavirus: On Crisis, Emergency, and the Power of Words." *Institute for National Security Studies*. No 1353. July 26. Web.

Schumpeter, Joseph. A. *Capitalism, Socialism and Democracy*. With a New Introduction by Richard Swedberg. London and New York: Routledge, 2003. Print.

Schwab, Klaus and Thierry Malleret. *Covid-19: The Great Reset*. Cologny/ Geneva: Forum Publishing, 2020. Print.

van Prooijen, Jan Willem. *The Psychology of Conspiracy Theories*. London and New York: Routledge, 2018. Print.

Is Klaus Schwab the Most Dangerous Man in the World? – YouTube.

Prince Charles is now selling his 'eco-fascist fantasy of the Great Reset' – Bing video.

The Great Reset: Joe Biden and the Rise of 21st-Century Fascism – YouTube.

#TheGreatReset – YouTube.

https://www.weforum.org.

Daniel Clinci

Posthumanism and Climate Change: Why Theory Matters

No science without theory
Nicholas Georgescu-Roegen

Abstract: In the face of ecological crises, scientific knowledge offers data and facts that must be interpreted and translated into political, social, and economic theory. A theory for the Anthropocene must take into account not only climate change, but also the whole underlying system of global domination and inequalities.

Introduction

In the first decades of the twenty-first century, environmental concerns brought about an unprecedented body of research on climate change and on the current state of the planet, the latest stage of the Anthropocene. Together with the scientific results and reports of various organizations, responses, approaches, and theories have emerged to fulfill the need to organize this knowledge into some kind of coherent future praxis. If anything, this proves the fact that a theory for the Anthropocene is very much necessary, a theory that can provide a framework not only from the points of view of philosophy or ontology, but also from those of politics, society, and economics. In this sense, theory matters, even if it seems to be currently exiled to the margins of both academia and activism; furthermore, critical theory matters as the indispensable tool in the process of re-thinking the world.

The most recent report from the Intergovernmental Panel on Climate Change (IPCC 2021) paints a grim picture in which temperatures have been steadily rising in the last decades as a result of "human influence," leading to melting glaciers and ice sheets, more frequent heat waves, and devastating droughts. The report also mentions that if business were to carry on as usual, an increase exceeding 2°C is to be expected during this century, reaching a global temperature which was last sustained over three million years ago (IPCC 17). On the other hand, the latest report from the Intergovernmental Science-Policy Platform on Biodiversity and Ecosystem Services (IPBES 2019) focuses on "nature" and notices that what we might refer to as "extraction" (deforestation, mining, industrial fishing etc.) has increased since 1970, while one million species of animals and plants face extinction in the coming decade (IPBES XV). The report also lists a number of factors which led to the current situation, such as overexploitation, climate change, global economic growth (a fourfold increase in the past five decades), and global trade (a tenfold increase in the past five decades) (IPBES XVI–XVIII). It also mentions that these factors are unequally distributed (IPBES XIV); for instance, even if there is enough agricultural output for global needs, 11 percent of the world's population still suffers from undernourishment and starvation. A World Wildlife Fund report from 2020 shows that biodiversity has been steadily decreasing by a factor of 68 percent since 1970 (WWF 16). Thus, it is difficult to isolate climate change from the other issues that make up the Anthropocene; in their original article which established the term "Anthropocene" as a potential denominator for the current epoch, Crutzen and Stoermer begin indeed from the fact that the increase in "greenhouse gases" (carbon dioxide and methane, among others) is the significant factor in defining this new geological period (Crutzen and Stoermer 17). However, they also document many other anthropogenic effects "to emphasize the central role of mankind in geology and ecology" (Crutzen and Stoermer 17).

Considering these issues, in 2015, the Paris Agreement, a legally binding international treaty, was promoted to create the framework for the global mitigation of climate change, but also with targets such as eradication of poverty and sustainable development. The Agreement proposed a sort of global intergovernmental cooperation, focusing on, for instance, the transfer of technology, the importance of education and public participation, and the regulation of carbon sinks (forests).

However, a 2019 report from a non-governmental organization (FEU-US), shows not only that there has been no improvement in the amount of global greenhouse gas emissions, but there has actually been an increase, and international support has been lacking. In this context, there is little hope that the Paris Agreement will manage to fulfill its target of limiting global warming to a maximum of 2°C above preindustrial levels by mid-twenty-first century. In truth, the Paris Agreement was never intended to act as theory for future praxes, but as a mechanism for states to make their pledges and to be reviewed by their peers. As one paper shows (Barrett and Dannenburg), this mechanism is highly unlikely to produce the desired effects.

Mapping the horizon

In 2017, Donald J. Trump, then President of the United States, decided to withdraw from the Paris Agreement, voicing a much wider American public disinterest in environmental issues like climate change. The decision came into full effect in 2020, but the American apathy in these matters dates well back into the 1990s. For instance, in 1999, the percentage of Americans who treated climate science with skepticism was among the highest in the world (Norgaard 179). Ten years later, in 2009, Americans were once again found to believe that dealing with environmental problems should be among the last of their government's concerns (Norgaard 178). This is one of the social and political responses to the current climate crisis: skepticism and denialism. Norgaard identifies the seemingly paradoxical fact that the United States scientists have produced a great deal of knowledge regarding the climate, yet the American public and media treat it as "background noise" (Norgaard 183).

However, denialism and skepticism are, in fact, politically and socially engaged approaches – conservatism. Holding on to arguments such as the need to protect a fabled "American way of life," conservative think-tanks and corporations funded campaigns to silence any environmental concerns and to attack the scientific foundations of environmentalism (Jacques, Dunlap, and Freeman 352). This tactic allows the conservative movement to continue advocating for a reduction in government regulations and corporate liability, which they see as progress (Jacques, Dunlap, and Freeman 354). In other words, conservatives understand the opposition between neoliberal capitalism and the scientific discourse regarding the Anthropocene; as a result, they employ strategies based on questioning the validity of the knowledge provided by science, resorting to American exceptionalism and individualism (Norgaard 192).

Another contributor to the development of climate change skepticism is corporate advertising, especially that of fossil fuels corporations. As one recent study shows, ExxonMobil's strategy has a contradictory aspect; in internal documents and peer-reviewed publications, ExxonMobil acknowledges the fact that climate change exists and constitutes a major global issue. However, in advertorials and other further-reaching publications from *The Washington Post* and *New York Times*, the corporation's position is that of doubt, using the same tactics of questioning the scientific reports and arguing in favor of politics over science (Supran and Oreskes, *Addendum* 6).

However, there is another way out of the entire climate change debate for neoliberal capitalism, that is, individual responsibility. All the reports cited above, and many others, refer to climate change and to all the processes set in motion in the Anthropocene in terms like "anthropogenic," "human influence," "human-made," and so on, not accounting for the massive global inequality of the twenty-first century. Thus, the strategy of using that old concept of classical liberalism, "individual responsibility," is able to put the blame for climate change on the individual and to, firstly, construct the eco-conscious consumer; secondly, greenwash capitalism; and thirdly, promote "green" brands.

For instance, one of the ways in which the current narrative of "we are all to blame" for climate change was promoted and the eco-conscious consumer was constructed can be traced back to the same ExxonMobil advertorials, where another tactic emerged. Focusing on consumer demand and not on production, the corporation's publications construct a "Fossil Fuel Savior" frame (Supran and Oreskes, *Rhetoric*) to imply that rising greenhouse gas emissions due to increased demand is what drives climate change. The corporation eludes any liability by placing the blame on consumers. Thus, individual responsibility is called upon to solve climate change: efficient use of electricity and fuel, planting trees, carpooling etc. At the same time, the fossil fuel corporation markets itself as eco-conscious in what is now known as "greenwashing" by emphasizing their investment in the research and development of new, more energy-efficient technologies and by financing climate change studies at top universities (Supran and Oreskes, *Rhetoric*). The same paper lists other companies that use the same strategy of individual responsibility and greenwashing, such as packaging and beverage manufacturers, motor vehicles and leaded products producers, and plastic producers. Another example from a much more aggressive re-branding and marketing campaign is that of BP (formerly British Petroleum). In 2004, BP came up with the "carbon footprint calculator" to imply that everyday life is the main driver of climate change in an attempt to shift corporate liability to individual responsibility, as part of a larger greenwashing campaign (Doyle). Using the online calculator, one can measure the personal impact one has on the environment, reinforcing the concept of individual responsibility, while the corporation gains the moral high ground. The American supermarket chain Wal-Mart began to focus on turning "green" in 2005, pledging to become more efficient and reduce emissions. However, in the following year Wal-Mart's emissions rose by 9 percent, as the company was expanding in the United States and abroad (Foster, Clark, and York 388–389).

These examples and practices have an underlying political theory, in addition to remixing Adam Smith's "self-interest" with neoliberal approaches. Some authors have used the term "economic Malthusianism" to describe this strategy, echoing Thomas Malthus' late eighteenth-century ideas about population growth (Foster, Clark, and York 378).

Malthus' *Essay* (1798) deals with unchecked population growth which may cause potential food shortages; he also proposes two checks to maintain the balance between population and food (or resources): raising the death rate and lowering the birth rate. In the revised 1803 version, Malthus introduced another check, moral restraint, such as voluntary celibacy (Malthus 83). In the same way, the current economic Malthusianism proposes that consumption should be limited and controlled voluntarily, while corporations go about doing business as usual. However, the more traditional and more direct interpretation of Malthus is "demographic Malthusianism," that is, global population growth is responsible for climate change. Indeed, overpopulation is often included in the list of factors that make up the Anthropocene.

In 1968, Paul R. Ehrlich published *The Population Bomb*, a highly influential book that is literally the twentieth century Malthusian *Essay*. Ehrlich argues that overpopulation is a global issue that needs to be addressed beginning from America, because Americans consume a large part of global resources even though they make up only a small part of the global population (Ehrlich 129); at the same time, Ehrlich promotes the same individual responsibility, or Malthusian moral restraint, advising that one should have no more than two children (Ehrlich 159). Ehrlich notices early on that underdeveloped countries (two thirds of the global population, from Asia, Latin America, Africa) have rapid growth rates and overdeveloped ones have slower growth rates, and that the poorer countries start having "rising expectations" of affluence (Ehrlich 7–8). One solution Ehrlich proposes for solving this issue in the case of India is to employ forced sterilization of males with three or more children (Ehrlich 151).

This demographic Malthusianism, "population control in the Global South as a potential solution for climate change" (Dyett and Thomas 206) has led to the current socio-political approach known as eco-fascism, which can be seen in the public discourse of, for instance, former Vice-President of the United States, Al Gore, and The Bill and Melinda Gates Foundation (Dyett and Thomas 211). In 2019, both the Christchurch (New Zealand) and the El Paso (Texas) shooters, who targeted Muslims and Latinos, respectively, were motivated by the belief in a link between overpopulation and environmental issues, as well as by the fear of an immigrant "invasion" (Achenbach; Lawton).

If corporate neoliberal discourses went to great lengths (aided by consistent funding) to reinforce the ideology of personal responsibility, eco-fascism refers to entire populations and races. It may seem that there is a world of difference between Gore's concerns about birthrates in Africa and emphasis on reducing populations in the underdeveloped world, or the Gates Foundation's efforts to prove that "family planning" is a cost-effective measure for the development of countries in the Global South (Dyett and Thomas 211), and the Christchurch and El Paso mass murderers, but on a closer analysis their ideas are strikingly similar to Ehrlich's 1968 complaint that there are too many people in the underdeveloped world and that something must be done about it. Eco-fascism, either explicitly or not, translates the issue of overpopulation in racial terms, identifying the African, the Indian, the Latino, and so on as the post-colonial Other causing climate change.

Eco-fascism fails to consider the wider implications of the global system of inequalities; for instance, a report from one of the most important environmental organizations in the United States, the Sierra Club, makes a much more informed connection between population growth and consumption of resources, and notices that the USA is the most overpopulated country on the planet because it consumes the most resources per capita (Debes 1). Another report shows that North Americans consume almost nine times more resources than Africans (Friends of the Earth 20). Thus, the issue may not necessarily be overpopulation, but overconsumption.

In stark contrast to these theories and practices that place the responsibility for climate change and environmental crises either on the individual or on communities defined by race or religion, eco-socialism identifies the system of global capitalism as the driver of climate change. From the point of view of eco-socialism, tackling climate change means doing away with capitalism, an idea summarized in the widely used slogan "system change – not climate change" (Empson 6; Pettifor 31).

In the 1970s, ecological questions began to be analyzed in a more systemic manner in both the scientific discourse and theory, as James Lovelock published his groundbreaking "Gaia theory" and Lynn Margulis popularized her ideas about symbiosis. At the same time, critical theory and ecofeminism contributed to the process of dismantling power structures by revealing their underlying assumptions. For instance, István Mészáros' 1970 lecture "The Necessity of Social Control" included an explanation of how the American "high mass-consumption" (Mészáros 27) had been proposed as a universal way of life, not accounting for the fact that it would deplete global resources at an unprecedented rate. He also criticized demographic Malthusianism, identified capitalism as a "system of waste production," and noticed the fact that "human survival" actually referred to the survival of capitalism, a system bound to destroy the planet. In the same lecture, Mészáros recuperated some of Marx's ideas on ecology, insisting that social change must be global and must have a broad scope. However, throughout the 1980s, ecological thought and action remained a utopian "green" middle-class affair which oftentimes refused any association with the political left (Porritt 116).

Only in the 1990s will more coherent ideas be developed, when deep ecology met critical social theory; Félix Guattari argued for an "ecosophy" or a "generalized ecology" that would transcend mere environmental concerns and would act transversally as a counterattack on what he named "Integrated World Capitalism" (Guattari 52). However, Guattari's ideas did not immediately permeate much of the English-speaking discourse, being translated only in the 2000s.

In 1993, a significant and influential approach regarding eco-socialism was formulated by David Pepper, who discussed the possibility of conflating "green" concerns with an ecological aspect of the Marxist dialectic between man and nature, arguing that eco-socialism did not stop at criticizing capitalism, but also envisaged new ways beyond it (Pepper 232–234). In the 2000s, this move towards a Marxist ecology, or eco-socialism, was complemented by John Bellamy Foster's rediscovery of Marx's concept of "metabolic rift" (Foster 155). When Marx tackled the Malthusian overpopulation issue, he resorted to the research of soil chemist Justus von Liebig, showing that food production was dependent on soil fertility which, under the conditions of industrial capitalism, was dropping fast due to a number of factors, the most immediate of which was the fact that "soil metabolism" was broken. In order to ensure that the soil remained fertile, a certain reproduction of this resource had to occur. But due to urbanization and industrialization organic recycling was impeded. This led Marx to elaborate on his theory of the "metabolic rift": in capitalism, the soil is robbed of its fertility (Foster 155; Marx 352–353). In a similar way, since human lives and nature are dynamically intertwined in the same metabolism and ultimately labor and nature are the two combined sources of wealth, capitalism instrumentalizes both in order to make profit.

Eco-socialism, the understanding that economy, ecology, and society are deeply inter-related, has been the foundation for some of the recent Green New Deal proposals. However, in order to construct potential socio-economic solutions to climate change, one very difficult thing has to be done: to do away with that Thatcherist legacy which says that "there is no alternative" to capitalism, that is, to do away with capitalist realism (Fisher 2). In a sense, Green New Deals have been stuck between a radical, revolutionary form (presented by the initial proponents, in Britain) and a more immediate "green capitalism & social justice" form. There is a possibility that both are utopian; radical and systemic changes seem quite improbable if we accept Thatcher's mantra, but a greening of capitalism is almost an illusion.

The green deals

In the wake of the 2008 global financial crisis, various groups and institutions began to envisage systemic changes under the label "Green Deal." In Britain, a workgroup of environmental activists and economists devised a "Green New

Deal" (Pettifor); at the United Nations, a "Global Green New Deal" was proposed (Barbier); ten years later, in the United States and the European Union, versions of a "Green New Deal" started being debated.

    These proposals are very different, both in the way they approach the current crises and in their scope. The one that is closer to eco-socialism, even if the label is not uttered one single time in the book, perhaps strategically, is Ann Pettifor's British Green New Deal. Pettifor begins by arguing for "system change, not climate change" and by identifying the global financial system of neoliberal capitalism as the ultimate culprit for the current environmental disaster (Pettifor 22). She points out that financial deregulation on a global scale led to a pattern of unsustainable consumption and, in Marxist or eco-socialist fashion, explains that this system exploits both human beings and nature. Her main solution is state intervention, which only sounds radical because four decades of neoliberalism managed to completely divorce the state from the workings of the market. However, she also calls for global change (recognizing the fact that climate change, for instance, is a global phenomenon) towards a "steady-state economy" (Pettifor 70), that is, an economy which is no longer based on the neoliberal illusion of infinite growth, and which would be powered not by carbon, but by alternative sources and human labor. Thus, the Green New Deal would achieve full employment and would be able to cater to everyone's needs (Pettifor 75) without overexploitation.

    At roughly the same time, the United Nations commissioned Edward B. Barbier to elaborate a "Global Green New Deal" as an economic and political framework. Barbier's plan included a transition to renewable energies by using incentives and carbon taxes, which would create a large number of new jobs (Barbier 53); in keeping with other United Nations plans, he pointed out that reducing worldwide poverty could be achieved through "payments" for "ecosystem services," such as through afforestation (Barbier 126); development of fuel-efficient transportation was also included, with South Korea given as an example (Barbier 175). From a theoretical and political perspective, Barbier's plan lacks Pettifor's criticism of neoliberalism because it is an attempt to construct a tamed and ethical capitalism, a capitalism with a human face. In fact, the US Green New Deal proposed by Representative Alexandria Ocasio-Cortez and Senator Edward Markey resembles Barbier's United Nations plan and not Pettifor's version, in spite of the fact that the latter collaborated with Ocasio-Cortez.

    The Resolution is explicitly based on the IPCC's 2018 report, perpetuating the same language of "human"-made climate change; at the same time, it admits the fact that the United States has emitted 20 percent of historical greenhouse gas emissions and, consequently, it has the responsibility to lead the process of curbing these emissions. Also, the Resolution tends to think of greater systemic ills, such as racial and economic injustices and proposes the creation of unionized jobs, the secure access to clean air and water, the reparation of historic oppression of vulnerable communities (indigenous peoples, migrants, women etc.). Having a national or local focus, the Resolution promotes local manufacturing using renewable energy sources, the use of "green" transportation, and an end to the transfer of jobs overseas. As Max Ajl points out, this green social-democratic/green capitalist program is imperialist, because it still relies on resource extraction from underdeveloped countries, for instance, the lithium needed for batteries, (Ajl 95) without any plans of global wealth redistribution.

These potential political and economic interventions suffer from some major drawbacks. Firstly, they are interested more in keeping the lifestyles of the citizens of overdeveloped countries relatively intact; even Pettifor's more radical approach talks about a global steady-state economy without mentioning the horrendous inequalities that capitalism created and taking for granted that underdeveloped countries will have the means to employ such measures from a position of power equal to that of Western states. Barbier's text proposes a sort of eco-colonization – the Global South as a source of unskilled "ecosystem" labor for the Global North. Secondly, even the system change Pettifor is arguing for does not imply a change in social relations, even if she envisages a future where individualism and the principle of competition will be surpassed by cooperation. Social relations are always bound by their materiality, but Pettifor does not even dare to include the term "socialism" in her analysis. Thirdly, all these perspectives are anthropocentric, defeating any attempt to think and act beyond the Anthropocene.

Posthumanism as a new theory for the Anthropocene

Denialists, capitalists, either with a "green" veneer or downright "brown," neo-Malthusian eco-fascists, or the eco-politicians of the day begin from the fundamental assumption of Cartesian humanism – that the human and nature are two very different and even opposite things. Descartes framed this distinction as *res cogitans* (rationality, mind, soul) and *res extensa* (nature, inert matter), the foundation of all Western theory. The approaches discussed above still perpetuate this divide with various twists and turns, but the idea that nature is an instrument to be shaped and molded as humans please never leaves the background. In a sense, this is what the Anthropocene is all about, the "human epoch" which marks the domination of one species over the entire planet, leaving geological-stratigraphic traces that attest its power. If a theory for the Anthropocene is to be developed, then it must make nuanced observations about the historicity of this power, it must retrace the way it came to be, it must find alternatives, and it must also define this "Anthropos."

Deeply embedded in Western political, economic, and philosophical theory which have led to the current environmental crises is the idea of human exceptionalism. Posthumanism challenges these anthropocentric worldviews in multiple ways. For instance, a kind of Foucauldian analysis of the concept of humanism and the praxis of capitalism reveals the fact that they co-developed during the seventeenth and eighteenth centuries. For Hobbes (*Leviathan*) and Locke (*Second Discourse*), the human was defined as a combination of private property and labor. For the latter, God has given humans a world in common; through labor, humans conquer nature and thus acquire private property, from which they can derive a surplus (Locke 15). This concept of the human as labor and private property clearly shows that, historically speaking, not all members of the genus *Homo* were also "humans." The African slaves of the trans-Atlantic triangular trade mechanism, indigenous American peoples, Western European women and children, or the nascent proletariat did not fit the definition; most often, they were either reduced to labor power (having no property), or mere social decorations (having no property and no labor power). The "human" was actually the bourgeois, Christian, white male, whose legacy is perpetuated in the current global systems of oppression and inequalities in both consumption and pollution. Since nature was a gift from God, it became a moral duty to enclose or extract private property (Thompson 165).

In the 1970s, a range of new critiques and theories fundamental to the development of a posthumanist perspective began to appear in multiple domains. James Lovelock's concept of "Gaia" as the cybernetic, self-regulating, and self-sustaining Earth-system was the first to acknowledge that there is a deep interconnectedness of all life forms and geological processes. According to Lovelock, there is a systemic metabolic process on Earth that maintains life and adapts to any changes so as to preserve it. Commonly misinterpreted as "Mother Earth," Lovelock's Gaia is not only the sum of living beings on the planet, but also the atmosphere with which they interact in order to manipulate it for their needs. Whenever new conditions appear, such as too much carbon dioxide, the feedback loops create "adaptive changes" to compensate, including climate changes (Lovelock 8–9). One relevant point that Lovelock makes is that these changes may affect human populations, but they will be beneficial for life on Earth.

When Lynn Margulis developed her theory of symbiogenesis in the late 1960s, she did not think on a global level. Margulis was interested in how organisms work together to create new life forms; the first example is the mitochondrion, which was initially a bacterium (Margulis, *On the Origin*). Even though she had worked with Lovelock on the Gaia framework, it was only in the 1990s that she understood the fact that Gaia works by symbiosis or autopoiesis, that is, living beings work together to create the conditions for their own existence, as a whole, in a kind of closed-loop metabolism. According to Margulis, Gaia is a recycling machine – living beings produce waste that is used by other living beings. As a result, "Gaia is a tough bitch" and "anthropogenic" climate change will only affect human life, not all life (Margulis, *The Symbiotic Planet* 161).

But even more relevant for the development of a posthuman worldview in Lovelock and Margulis is the acknowledgment that life, including the genus *Homo*, and the environment are one single thing (Margulis, *Acquiring Genomes* 129). In the field of economics, these theories have been mirrored by Nicholas Georgescu-Roegen's work on ecology. Trying to introduce a thermodynamic understanding of the economic processes instead of the more traditional mechanical one (Georgescu-Roegen, *The Entropy* 276), Georgescu-Roegen was interested in, among other things, the production of waste, including "thermal pollution" (Georgescu-Roegen, *Energy* 13-14). He tried to prove, with a certain degree of pessimism, using the second law of thermodynamics, that economic activities necessarily increase entropy, that is, unavailable energy. For Georgescu-Roegen, this means that humankind will become extinct, because it will be unable to either return to a previous state or limit the extraction of resources for its own "exosomatic comfort."

As we can see, the theories mentioned above already challenge the traditional Western divide between "human" and "nature." However, the foundation of a more constructive posthumanism was created in the 1980s by the transdisciplinary approach of Haraway's "A Cyborg Manifesto," which tried to provide an intersection between socialism, feminism, and materialism (Haraway, *Manifestly* 5). Haraway treats the inherent dualisms of the Western tradition as strategies of power through which the dominant "autonomous self" (Haraway, *Manifestly* 60) subjects "women, people of color, nature, workers, animals" (Haraway, *Manifestly* 59). The ironic-mythical figure of the cyborg is an intersection of all these and, ultimately, a proposed escape from the capitalist-humanist systems of domination.

New feminist materialisms have continued this intersectional work by using the eco-legacy of the 1970s to show that bodies are co-constituted within flows of matter-energy (Alaimo's concept of "transcorporeality"; Bennett's "vibrant matter"). Haraway herself began using the concept of "naturecultures" to denote the *sym-biosis* (living together) or *sym-poiesis* (becoming-with) of species interactions, following Margulis (Haraway, *When Species* 32). In fact, new feminist materialisms argue that "we" have never been individuals, but always in-the-making as trans-species, matter-energy intersections, always posthuman.

Capitalism, on the other hand, commodifies, encloses, and draws boundaries (De Angelis 57; Cudworth and Hobden 153-154; Stengers 80-81), in other words, it segments (Deleuze and Guattari 210). As a result, the concept of a symbiotic, auto- or sympoietic Gaia radically opposes capitalism's necessary processes of territorialization, of eco-nomics as οἰκο-νόμος, law and management, and also as οἰκο-νομός, division. Thus, capitalism is an ecological (or ecosophical) regime of power, not only in the sense that matters for climate change, but also in the one that matters for the entire intersection of nature and culture, rigged to create waste.

The co-emergence of capitalism and humanism as the enclosure of rational self-interest and private property led to colonization and commodification, that is, to the transformation of anything into "resources." This "epic version of materialism" (Stengers 58) and the new posthumanist materialisms are the current political conflict, and climate change is at stake, because it is obvious that capitalism cannot "degrow," be "green," "decarbonized," or "sustainable"; it cannot function without perpetually extracting value from "resources." In spite of the recent trend towards a more ontological or fuzzy concept of posthumanism (Ferrando; Braidotti), going back to the fundamental ideas of the 1970s and 1980s is an essential part in configuring a political theory for the Anthropocene that is both critical and constructive, respecting the original intersectional framework of Haraway's work.

Eco-socialism, as we have already seen, provides important interventions in its re-reading of Marx in the context of climate change (Foster); new materialisms, powered by the scientific lineage of Lovelock and Margulis and the theoretical one of Deleuze and Guattari, challenge the old materialism of nineteenth century extraction capitalism (Alaimo; Bennett); new intersectional feminisms highlight the deep mechanisms of power at the heart of humanism-capitalism (Preciado; Cremin).

Scientific research, as we have claimed, shows that climate change is not a singular and nuclear issue, it is inter-connected within the entire world system, and that is why global organizations have called for wider social, economic, and political changes. However, apart from the fact that global agreements have failed to act decisively so far, it seems that resistance and critique remain the most immediate answers. Of course, they are fundamental in creating an alternative framework to think beyond the Anthropocene, but they are not sufficient. Also, it may be argued that global top-down solutions are not functional because the current world system is too dependent on extraction, production, and trade;

local top-down solutions suffer the same fate because they fail to understand the current system as *world* system.

A posthumanist theory for the Anthropocene begins from a low theoretical (Wark xvi) perspective and, understanding the entanglement of life forms, Earth processes, things, and technologies in their materiality, advocates for humility (Cudworth and Hobden 147). Posthumanism, or the flavor of posthumanism presented here, proposes cyborg communities against patriarchal humanist capitalism. In the words of Haraway, recognizing the materiality of our entanglement with the world suggests making-kin with co-constitutive ("sympoietic") "earthlings" (Haraway, *Staying* 102-103) as compost: "critters – human and not – become-with each other, compose and decompose each other" (Haraway, *Staying* 97). The task of a posthuman commune would thus be to dismantle the enclosures of capitalism at any level so that another trans-species politics can occur, one that minimizes domination (Cudworth and Hobden 147).

Conclusion

In 2014, a mathematical study analyzed the conditions in which societies collapse, factoring in social stratifications (Elites and Commoners) and the dynamics of production (Wealth and Nature). The authors found that social inequalities and overexploitation lead to total collapse; their model clearly shows that only egalitarian societies can survive in the long run (Motesharrei, Rivas, and Kalnay 100-101), proving that resilience can be achieved communally. Trying to preserve capitalism under the guise of a new corporate lexis which includes "sustainable development," "eco-friendliness," and "greenness" as its keywords defeats any communal and egalitarian alternative and advances the Anthropocene. Also, a 2008 study shows that climate change due to carbon dioxide accumulation in the atmosphere is irreversible (Solomon et al.), making Haraway's suggestion of "staying with the trouble" even more relevant. Perhaps a posthumanist theory for the Anthropocene may also include living-with climate change as a historical and material reminder of the human epoch.

It remains to be seen whether Georgescu-Roegen's pessimistic view will prevail, and societies will collapse under the weight of an unjust system, or they will re-frame and re-construct the trans-species communities that we have always been.

References

Achenbach, Joel. "Two Mass Killings a World Apart Share a Common Theme: 'ecofascism.'" *The Washington Post.* August 18, 2019. Web.

Ajl, Max. *A People's Green New Deal.* London, UK: Pluto Press, 2021. Print.

Alaimo, Stacy. *Exposed. Environmental Politics and Pleasures in Posthuman Times.* Minneapolis and London: University of Minnesota Press, 2016. Print.

Barbier, Edward B. *A Global Green New Deal. Rethinking the Economic Recovery.* Cambridge, UK: Cambridge University Press, 2010. Print.

Barrett, Scott and Astrid Dannenburg. "An Experimental Investigation into 'pledge and review' in Climate Negotiations." *Climatic Change* 138 (2016): 339–351. Web.

Bennet, Jane. *Vibrant Matter. A Political Ecology of Things.* Durham and London: Duke University Press, 2010. Print.

Braidotti, Rosi. *The Posthuman.* Cambridge, UK: Polity Press, 2013. Print.

Crutzen, Paul J. and Eugene F. Stoermer. "The 'Anthropocene.'" *Global Change Newsletter.* The International Geosphere-Biosphere Program of the International Council for Science. 41 (May 2000): 17–18. Web.

Cudworth, Erika and Stephen Hobden. *The Emancipatory Project of Posthumanism.* London and New York: Routledge, 2018. Print.

De Angelis, Massimo. "Separating the Doing and the Deed: Capital and the Continuous Character of Enclosures." *Historical Materialism* 12 (2004): 57–87. Web.

Debes, Peter. "Population." *Sierra Club Eco-logue* 46 (Dec 2016–Feb 2017): 1. Web.

Deleuze, Gilles and Félix Guattari. [1980] *A Thousand Plateaus. Capitalism and Schizophrenia.* Translation and foreword by Brian Massumi. Minneapolis and London: Minnesota University Press, 1987. Print.

Doyle, Julie. "Where Has All the Oil Gone? BP Branding and the Discursive Elimination of Climate Change Risk." *Culture, Environment, and Ecopolitics.* Edited by N. Heffernan and D. Wragg. Cambridge, UK: Cambridge Scholars Publishing, 2011. Web.

Dyett, Jordan and Cassidy Thomas. "Overpopulation Discourse: Patriarchy, Racism, and the Specter of Ecofascism." *Perspectives on Global Development and Technology* 18 (2019): 205–224. Web.

Ehrlich, Paul R. *The Population Bomb.* Revised. New York: Ballantine Books, 1988 [1968]. Print.

Empson, Martin. "Why We Need System Change." *System Change not Climate Change. A Revolutionary Response to Environmental Crisis.* Edited by Martin Empson. London: Bookmarks Publications, 2019. Print.

Ferrando, Francesca. *Philosophical Posthumanism*. London, UK: Bloomsbury, 2019. Print.

FEU-US. *The Truth Behind the Climate Pledges*. [Robert Watson, James J. McCarthy, Pablo Canziani, Nebojsa Nakicenovic, Liliana Hisas]. 2019. Web.

Fisher, Mark. *Capitalist Realism. Is There No Alternative?* Winchester, UK: Zero Books, 2009. Print.

Foster, John Bellamy, Brett Clark, and Richard York. *The Ecological Rift. Capitalism's War on the Earth*. New York: Monthly Review Press, 2010. Print.

Foster, John Bellamy. *Marx's Ecology. Materialism and Nature*. New York: Monthly Review Press, 2000. Print.

Friends of the Earth. *Overconsumption? Our Use of the World's Natural Resources*. 2009. Web

Georgescu-Roegen, Nicholas. *Energy and Economic Myths. Institutional and Analytical Economic Essays*. New York: Pergamon Press, 1976. Print.

Georgescu-Roegen, Nicholas. *The Entropy Law and the Economic Process*. Cambridge, MA: Harvard University Press, 1971. Print.

Guattari, Félix. [1989] *The Three Ecologies*. Translated by Ian Pindar and Paul Sutton. London and New Brunswick: The Athlone Press, 2000. Print.

Haraway, Donna J. *Manifestly Haraway*. Minneapolis and London: University of Minnesota Press, 2016. Print.

Haraway, Donna J. *Staying with the Trouble. Making Kin in the Chthulucene*. Durham and London: Duke University Press, 2016. Print.

Haraway, Donna J. *When Species Meet*. Minneapolis and London: University of Minnesota Press, 2008. Print.

IPBES. *Global Assessment Report of the Intergovernmental Science-Policy Platform on Biodiversity and Ecosystem Services*, Brondízio, E. S., Settele, J., Díaz, S., Ngo, H. T. (eds.). IPBES secretariat, Bonn, Germany, 2019. Web.

IPCC. "2021: Summary for Policymakers." *Climate Change 2021: The Physical Science Basis. Contribution of Working Group I to the Sixth Assessment Report of the Intergovernmental Panel on Climate Change* [Masson-Delmotte, V., P. Zhai, A. Pirani, S. L. Connors, C. Péan, S. Berger, N. Caud, Y. Chen, L. Goldfarb, M. I. Gomis, M. Huang, K. Leitzell, E. Lonnoy, J.B.R. Matthews, T. K. Maycock, T. Waterfield, O. Yelekçi, R. Yu and B. Zhou (eds.)]. Cambridge University Press, 2021. Web.

Jacques, Peter J., Riley E. Dunlap, and Mark Freeman. "The Organization of Denial: Conservative Think Tanks and Environmental Skepticism." *Environmental Politics* 17 (2008): 349–385. Web.

Lawton, G. (2019). "The rise of real eco-fascism." *New Scientist* 243 (2019): 24. Web.

Locke, John. [1689] *Second Treatise of Government and A Letter Concerning Toleration*. Edited with an introduction and notes by Mark Goldie. Oxford, UK: Oxford University Press, 2016. Print.

Lovelock, James. [1979] *Gaia. A New Look at Life on Earth*. Oxford, UK: Oxford University Press, 2016. Print.

Malthus, Thomas-Robert. *An Essay on the Principle of Population*. 1803 Edition. Edited and with an Introduction by Shannon C. Stimson. New Haven: Yale University Press, 2018. Print.

Margulis (Sagan), Lynn. "On the Origin of Mitosing Cells." *Journal of Theoretical Biology* 14 (1967): 225–274. Web.

Margulis, Lynn, and Dorion Sagan. *Acquiring Genomes. A Theory of the Origin of Species*. New York: Basic Books, 2002. Print.

Margulis, Lynn. [1998] *The Symbiotic Planet. A New Look at Evolution*. London, UK: Phoenix, 1999. Print.

Marx, Karl. [1867] *Capital. A Critical Analysis of Capitalist Production*. Translated by Samuel Moore and Edward Aveling, with an Introduction by Mark G. Spencer. Hertfordshire, UK: Wordsworth Editions, 2013. Print.

Mészáros, István. *The Necessity of Social Control*. New York: Monthly Review Press, 2015. Print.

Motesharrei, Safa, Jorge Rivas, and Eugenia Kalnay. "Human and nature dynamics (HANDY): Modeling inequality and the use of resources in the collapse or sustainability of societies." *Ecological Economics* 101 (2014): 90–102. Web.

Noorgard, Kari Marie. *Living in Denial. Climate Change, Emotions, and Everyday Life*. Cambridge, MA: The MIT Press, 2011. Print.

Pepper, David. *Eco-socialism. From Deep Ecology to Social Justice*. London and New York: Routledge, 1993. Print.

Pettifor, Ann. *The Case for the Green New Deal*. London: Verso, 2019. Print.

Porritt, Jonathon. *Seeing Green. The Politics of Ecology Explained*. Oxford, UK: Basil Blackwell, 1984. Print.

Solomon, Susan, Gian-Kasper Plattner, Reto Knutti, and Pierre Friedlingstein. "Irreversible climate change due to carbon dioxide emissions." *PNAS [Proceedings of the National Academy of Sciences of the United States of America]* 106 (2009): 1704–1709. Web. January 5, 2022. Web.

Stengers, Isabelle. [2009] *In Catastrophic Times: Resisting the Coming Barbarism*. Translated by Andrew Goffey. Open Humanities Press/meson press, 2015. Print.

Supran, Geoffrey and Naomi Oreskes. "Addendum to 'Assessing ExxonMobil's Climate Change Communications (1977–2014)'." *Environmental Research Letters* 15 (2020): 1–18. Web.

Supran, Geoffrey and Naomi Oreskes. "Rhetoric and frame analysis of ExxonMobil's climate change communications." *One Earth* 4 (2021): 696–719. Web.

Thompson, E. P. [1991] *Customs in Common*. London, UK: Penguin, 1993. Print.

Wark, McKenzie. *Molecular Red. Theory for the Anthropocene*. London and New York: Verso, 2015. Print.

WWF. *Living Planet Report 2020 – Bending the curve of biodiversity loss*. Almond, R.E.A., Grooten M. and Petersen, T. (eds.). WWF, Gland, Switzerland, 2020. Web.

Roger L. Nichols

American Indians and Developing Eco-Consciousness

Abstract: During the last half century American Indians have demonstrated against federal and state actions that threatened their cultural and religious sites. The protesters sought help from environmental groups and their combined efforts energized public interest in ecological issues. My discussion examines how Indian actions to protect sacred sites benefited from and affected the environmental movement.

Introduction

Despite the fact that the 17th century European invaders thought that the Indigenous people in North America had no culture or religion, the latter had well-developed cultural and religious systems. The newcomers' ignorance of that led to policies and actions that brought disputes from then to the present. During the twentieth and 21st centuries American Indians became more assertive, demanding that their cultural practices and religious rites be recognized as legitimate. Those demands led to disputes with federal agencies overseeing much of the federal land in the West, particularly when federally approved actions interfered with tribal religious sites or rites. The arguments focused on a range of mostly federal actions related to religious sites. These included building a ski lodge and astronomy observatories on sacred mountains, copper mining and rock climbing on others, flooding reservation lands with water from hydroelectric dams, and an almost total rejection of Indian religious rights. When the U.S. rejected indigenous claims for religious freedom, tribal people shifted to making environmental arguments attracting sympathy and wide support from ecologically aware organizations and national religious organizations. Combining these two elements in their protests shows how that brought allies who otherwise might not have supported tribal demonstrations based on purely religious grounds. At the same time their protests brought national attention and growing support to the environmental movement.

Since as far back as the 1930s American Indians have participated in public demonstrations, marches, and confrontations related to environmental issues in places as far apart as California and New York, Wyoming and Arizona. Often the media presented these actions as protests against federal and state policies or acts. While partly true, much of this activity focused on positive assertions of indigenous sovereignty, not always negative attacks on the government. Basing their claims on long-existing treaty rights to land and resources, their beliefs and uses of sacred sites, and their ecological concerns, they objected to a variety of federal, state, and local actions and situations. Their repeated demonstrations and court suits kept environmental issues in the news for much of the past eighty years and added to the growing ecological awareness developing in American society.

To express their ideas and raise public awareness they used three broad sets of tactics. The first was physical confrontation. This included occupations of public spaces and buildings, large demonstrations, marches, and cross-country caravans. A second closely related tactic involved making proclamations and statements to the media to attract maximum attention, understanding, and support. Their third strategy used law-suits hoping to gain favorable rulings through the courts. Although their demonstrations and statements attracted the most public attention, the lawsuits, even when not always successful, often had a more significant impact than the other strategies. Whichever tactic or variety of actions they used, their efforts focused on broad questions related to closely intertwined environmental matters, religious issues, and Indigenous treaty rights.

Indians and the environment

For many reasons Americans have come to associate Indian people with the natural environment. The 1971 image of Iron Eyes Cody as the "Crying Indian" made the most direct connection between American thinking about ecology, the natural environment, and Indians. It is likely to have had much the same impact as the 1941 image of Smokey the Bear had in drawing attention to the dangers of forest fires. The Crying Indian picture first appeared on a poster entitled "Get Involved Now. Pollution Hurts All of Us" in connection with Earth Day activities. Then eight years of television public service announcements followed. Produced by the Keep America Beautiful organization, the nation's largest community improvement group, it depicted Cody weeping as he looked at trash floating in a polluted stream. The powerful television image of tears running down Cody's cheeks put Indians at the center of growing environmental concerns in American society. There can be little doubt that the Crying Indian image did more than anything else to cement the image of the American indigenous people to the growing environmental movement. However, it was not alone. It stood upon the shoulders of Indian campaigns involving environmental questions that stretched back for several generations. Popular connections between Indians and nature in European and American cultures had a long history, and for generations writers had depicted indigenous people as Noble Savages living as part of nature. (Kretch 15–28, Berkhofer 1978 and Pearce 1953).

Defending religious sites

One variety of indigenous activism resulted directly from religious beliefs and practices related to the physical environment. Because these differed so widely from the European invaders' Christianity, the whites ignored them. Within the context of settler colonialism, they rejected Indigenous beliefs tied to any specific natural features such as lakes, forests, mountains, or local natural sites. Despite this, since the 1978 passage of the American Indian Religious Freedom Act many Native groups assumed incorrectly that the law would help them protect their holy sites. That has not been the case, and most tribal protests based on religious practices tied to the environment have failed. One of the longest disputes over sacred sites began in the late 1930s and is related to building the Snow Bowl Ski Resort in the San Francisco Peaks outside Flagstaff, Arizona.

According to the Navajo people's creation story their ancestors emerged into the world at the San Francisco Peaks. Along with eleven other tribes in the area, they and the nearby Hopi consider the peaks as sacred. For generations they have used them for religious rites, and as a source of plants used in healing ceremonies. (Protect the Peaks, 5). In 1898 President William McKinley established the San Francisco Mountains Forest Reserve. Ten years later it became part of the new Coconino National Forest. In 1915 the situation changed when early Scandinavian immigrants to Flagstaff, only seven miles away, began skiing on the peaks. Two decades later, in 1938, what became the present Snowbowl Ski Resort began operations there over tribal objections. From its crude beginnings, it expanded to its present size, and now covers 777 acres (325 hectares) having multiple ski lifts, over 50 miles (82 km) of trails, and a paved road leading to shops, restaurants, and visitors' lodges. All of this operated under a forty-year special use permit renewed by the U.S. Forest Service in December 1980 (Navajo Nation v. U.S. Forest Service 2–4).

Responding to the coming expansion of Snowbowl facilities, the Navajo Medicine Men's Association filed a suit to block that move in the District Court for the District of Columbia. They sought a gradual removal of the skiing facilities or, if that failed, they wanted to get an injunction against any further expansion, alleging that the planned growth would "violate the Indians' first amendment right to the free exercise of their religion" by disrupting the natural environment on the peaks. This began a series of court suits which kept environmental matters in public view for the next thirty years. (Finnerty 6–7). In 1982 the district court ruled against the Indians on all counts. When they appealed, the Court of Appeals also ruled against them. It found that the resort did not impair their right to free exercise of their religion, that the Snowbowl area was not indispensable to their religious practices. They could use other parts of the mountains (Smith 169–171).

The decision halted Native objections temporarily, but in 2002 the dispute added a clearly environmental dimension. That year the City of Flagstaff offered to sell reclaimed sewage water to the resort for use in making artificial snow. In February 2005 the U. S. Forest Service approved the proposal and, a few months later, six tribes joined by the Sierra Club, the Southwest Center for Biological Diversity, and the Flagstaff Activist Network filed suit in the *Navajo Nation v. United States Forest Service,* to block that action. (Benally and Goodman 5). The district court denied the tribal claims supporting the Forest Service. When the Indians appealed, the Circuit Court upheld the earlier decision. Later the U.S. Supreme Court refused to consider the tribes' appeal. In 2011 construction of pipelines to carry the wastewater to the resort began and whites joined Indians as they tried to disrupt the work by chaining themselves to the equipment. The dispute went back into federal court with the decision ruling in favor of the resort again. In 2012–13 the Snowbowl began using the recycled water over continuing protests by Indians and environmentalists. Today it is the only ski resort in the world to use 100% wastewater for its snow making (Protect the Peaks, FACTS 1).

This dispute led to the creation of the Save the Peaks Coalition and a second group Protect the Peaks as a direct result of the Navajo and Hopi religious and environmental concerns. As a result, in northern Arizona the repeated tribal objections to the actions of the City of Flagstaff, the Snow Bowl Resort, and the U.S. Forest Service have kept ecological matters in the public eye to the near present. After losing repeatedly in federal courts the tribal suit reached the Arizona State Supreme Court. It too ruled against the Hopi's objections to the resort's use of treated effluent sewer water for its snowmaking. However, several of the justices accepted the Indians' connections between the environment and their religious practices. In his dissent to the decision Justice Scott Bales noted that "in the spring melt …. The myriad chemicals in the water will wreak unknown damage on the local ecosystem" (*Arizona Daily Star,* 2 Dec. 2018, C3 print).

Controversy over another Arizona mountain followed a different path. In 1984 the University of Arizona chose Mt. Graham in southeastern Arizona as the site for a new international astronomical observatory. It was to have 18 new scopes, and the imposing list of cooperating institutions included the Max Plank Institute, the Vatican, as well as Harvard University, the University of Minnesota, Notre Dame, Ohio State University, the University of Virginia, and the Smithsonian Institution (Helfrich, Metzger, and Nixon 2). In their enthusiasm for the project, University of Arizona leaders overlooked two crucial facts. First, Mt. Graham's three peaks were the home of an endangered species of red squirrels as well as several stands of the oldest conifer trees in the United States. (Stromberg and Patten 89; Swetnam and Brown 24–28) Second, the Apache people have considered the mountain a sacred place for at least five hundred years. Ignoring or perhaps ignorant of Apache beliefs, university officials got Congress to pass legislation that allowed them to build the first three of the telescopes while ignoring both the provisions of the American Indian Religious Freedom Act and the need for an environmental review. (Dougherty, 5) In response to the university's actions in 1989 the tribe organized the Apache Survival Coalition, declared the mountain a sacred site council, and organized public protests. (Helfrich 151–175)

Apache opposition drew wide public attention. In 1994 this led twenty-one environmental groups and individuals to establish the Mount Graham Coalition. It filed a suit asking that the university be required with the environmental laws (Dougherty 9). Their efforts brought increasing media attention and persuaded most of the universities to withdraw, leaving only Notre Dame and the Vatican as partners in the project. The Vatican, however, supported the observatory strongly. It sent Jesuit Fr. George Coyhne, director of its observatory, to examine the site. He claimed that he could not find any Apache who considered Mt. Graham as sacred. Then, after a brief climb, he reported seeing no shrines or other religious structures there. Sounding like a 16th century priest with the Conquistadores in Mexico, he dismissed Apache beliefs and practices as "a kind of environmentalism and religiosity to which I cannot subscribe, and which must be suppressed with all the force we can muster" (Mount Graham 2; Brandt 11). At that point the University of Arizona added to this stunning cultural insensitivity by naming the newest scope on the mountain the "Columbus."

Continuing Apache objections to the observatory gained wide public attention and broad support from national environmental and religious groups. The National Council of Churches endorsed the tribal claims and, in 1995, called for the removal of the telescopes from Mt. Graham (Mount Graham 2; Brandt 11). The President's Advisory Council on Historic Preservation declared that the telescope project violated the National Historic Preservation Act and, in 1997, President Clinton vetoed a $10 million appropriation for the university to operate the observatory (Brandt 12, 14; Mount Graham 3) The Apache filed suit in federal court but lost. In the meantime, while environmental studies show a continuing drop in the red squirrel population on the mountain, the university now requires all Indians wanting to visit the mountain for prayer to get a permit (Mount Graham 3).

A third example of the centrality of physical sites to Indian environmental views and religious practices occurred at the Devil's Tower National Monument in Wyoming. For centuries this 867foot (263 m) set of rock columns has been a sacred site for more than twenty Indian groups whose members come to it for prayer, vision quests, and to hold Sun Dances. At the same time the monument has become a cherished site for thousands of rock climbers whose actions there angered the Indians. As the controversy became heated both sides turned to environmental arguments. The Indians complained that the whites left their climbing bolts anchored in the rocks, stained the Tower's surface with urine, and scattered paper trash wherever they went. In response, one climber voiced the ideas of many when he described the Indian prayer bundles as "a bunch of trash… hanging around the monument." He went on to say that "the Indians don't climb that rock that I own as an American Citizen" (Dustin and Schneider 2). An Indian responded to that criticism with "all hell would break loose if I went to climbing the Washington Monument or the National Cathedral" (Dustin and Schneider 2).

The dispute placed the National Park Service, which administered the site, between the two groups and, in the early 1990s, it prepared a climbing management plan to ease tensions. The arguments centered on the fact that during each June summer solstice when the Indians held their sacred ceremonies at the Tower the number of rock climbers rose dramatically. Park Service officials at the site proposed a total halt to climbing in June, but tribal leaders demurred. They hoped to persuade the climbers to avoid the Tower in June out of respect for Indian religious practices. So, the Park Service persuaded individual climbers to accept voluntary ban to recognize and honor Tower's significance for Indian religious practices. At the same time, it banned commercial rock-climbing businesses from operating during June. In 1995 this idea went into effect (Indian Law Resource Center 1).

This new policy achieved a temporary drop of 84% in climbing activities during June, but failed to end the controversy. In early 1996 the Mountain States Legal Foundation filed a legal challenge to the agreement for the commercial rock-climbing owners whose businesses halted during June. Their suit charged that the Park Service initiated compromise violated the first amendment to the U.S. Constitution separating church and state. For the next several years the cases worked their way through the court system until 2000 when the U.S. Supreme Court ruled that the compromise did not favor or inhibit religious activity (Indian Law & Resource Center 1; Dustin and Schneider 3). At the same time the Court forced the Park Service to end its ban on offering June climbing tours (Collins 258. The dispute resulted from white rather than Indian legal action and illustrates how Indian religious practices that focused on specific natural sites kept environmental questions before the public, whether they won or lost their disputes.

Defending tribal lands

Indigenous peoples saw not only their religious connections to their environment being ignored or undercut by government actions, but their physical ones as well. Since the 1940s federal hydro-electric projects have reduced tribal land areas, flooded agricultural and forest land, long used burial grounds, and forced reservation dwellers to relocate their homes, farms, and ranches. Working through the U.S. Army Corps of Engineers the federal government authorized flood control and hydro-electrical watershed projects on major streams across the country. In 1957 they completed the Dalles Dam on the Columbia River in Washington. Five years later when it began operations in 1962 the Oahe Dam created an artificial reservoir that stretched 231 miles (372 km) on the Missouri River. What became Lake Oahe lay behind the second largest earthen dam in the world. It flooded nearly 161,000 acres of tribal range and cropland, as well as most of the standing timber on the Standing Rock and Cheyenne River Sioux reservations. Clearly this dam had a major negative impact on the environment on the Upper Missouri River Valley watershed (Shanks 573–579; Lawson 50–52).

Despite the ruinous impact these dams had on Indian groups living along that river, their complaints brought little redress and only modest public awareness. However, a third flood control dam built on the Allegheny River in northern Pennsylvania played a more prominent role in the impact Indians had on public awareness of water-related environmental issues. The Kinzua Dam built on the Allegheny River in northern Pennsylvania offers another example of the Corps of Engineers transforming Indian waters and lives. The Engineers created Kinzua Dam and Lake which received far more public attention than the Missouri River structures had. Much smaller than Lake Oahe on the Missouri River, when completed in 1965, the lake stretched only 25 miles (320km) up the river. The Senecas call it Lake Perfidy [Liars Lake], or the Lake of Betrayal.

Supposedly the dam came about in response to disastrous floods on the Allegheny River in 1928, 1936, 1938, and 1941 (Josephy 4). Those massive overflows threatened Pittsburgh and other cities in the upper Ohio Valley, and they asked for relief. But as Laurence Hauptman points out, manufactures in Pittsburgh and the upper Ohio River Valley needed a dependable source of electricity more than they needed flood protection (Hauptman 49–50). Moving quickly, in 1941, the Engineers proposed several flood-control dams, and soon began surveying on the Seneca Reservation in southwestern New York. About 1,800 Senecas lived on a mile wide strip of land on either side of the river that stretched 42 miles along the Allegheny River. They had lived there for generations and assumed that the Treaty signed by President Washington and tribal chief The Corn Planter in 1794 protected their land, but that agreement failed to save their homeland. In 1956 the Corps of engineers told the Indians that Congress had appropriated funds for building Kinzua Dam (Josephy 1–3).

To build it the Corps condemned 10,000 acres of tribal land. When completed, the dam left only 2,300 acres of useable land for Seneca farmers. Their experience resembled that of the Sioux in South Dakota. The dam would flood most of the tribes' arable land. It also forced one third of the reservation dwellers to relocate. The new lake covered long used tribal graveyards, and 3,000 sets of remains had to be moved. It also covered the small Cornplanter Tract which the State of Pennsylvania had given to the chief after the American Revolution for his help during that war (Armas and Hoover 10).

All of this happened after the Engineers persuaded Congress to fund the project at hearings, while they never invited the tribal leaders to attend. Once the Seneca learned about the Dam's possible destruction of their homeland they protested. They called for an injunction to keep the engineers off tribal land, but the courts rejected their claim. Then, accepting the need for some flood control, they hired prominent engineers, Dr. Arthur Morgan and Barton Jones to help. Morgan had served as the chairman of the Tennessee Valley Authority, the largest and most successful flood control organization in the country. Jones had built the Norris Dam, a major part of the TVA project. These two men developed an alternative plan. They proposed diverting the flood water north into Lake Erie for less money and without inundating the Reservation. When they presented their plan to Congress the Corps and its many supporters testified against their alternative, so it failed (Josephy 4–5).

In 1961 Seneca leaders appealed to President Kennedy, asking that he use his authority to block the funding for the dam. However, their pleas came at the height of the cold war and the perceived need to keep the nation's industrial base strong. As a result, he refused to stop the dam plan. That was no surprise, as he had already approved earlier dam projects—and had personally dedicated the Oahe Dam on the Missouri. On August 9, 1961 he told Seneca leaders "I have now had an opportunity to review the subject and have concluded that it is not possible to halt the construction of the Kinzua Dam…Impounding the funds appropriated by Congress…would not be proper." (Kennedy web). Three years later the tribe lost another appeal when it objected to New York State's plan to build a new interstate highway directly through the reservation. The new road cut the Seneca reservation directly in half and made it impossible for families and friends to cross the road to visit. This latest assault on their land persuaded many of the Indians living on the much smaller reservation to move into nearby towns (Hauptman 90–100).

While the Seneca had held fewer and smaller public protests about the federal and state actions than either the demonstrators at the Dalles or Oahe Dams, they attracted some more national attention. Perhaps that came because some whites had to leave their small towns being flooded by the reservoir. During the controversy over building the dam Oliver LaFarge wrote a protest song entitled "As Long as the Grass Shall Grow." In it he denounced the dam and cited President Washington's promise to Chief Corn Planter that the Senecas would be able to live on the now flooded land in perpetuity. Country Western star singer Johnny Cash put this song first in his 1964 album *Bitter Tears: Ballads of the American Indian*. Canadian Cree folk singer Buffy Sainte-Marie devoted the fourth verse of her popular ballad "Now That The Buffalo's Gone" to the Seneca experience with the Kinzua Dam. There she points out that the treaty signed by George Washington and Seneca Chief Cornplanter is "being broken by the Kinzua Dam" (Sainte-Marie157). The protests continued to attract public attention to the near present when in late 2017 the Public Broadcasting System television showed a documentary film "Lake of Betrayal," about the Kinzua Dam (Toward Castle Films).

Seneca protests and legal efforts brought some federal responses. The government gave the tribe 305 acres of land in two nearby Seneca towns to resettle individual families. In 1964 Congress appropriated 15 million dollars in reparations for the tribe's lost land. It made other funds available for relocating 3,000 Indian graves to two new cemeteries, to develop an industrial park, and a 1.8 million fund for scholarships to college, business and vocational school for Indian students. Still, whatever the government offered each of the tribes could not heal the trauma of losing homes, cemeteries, sacred sites, and fertile land. Now tribal members look out on a vast lake filled with the boats of vacationing whites rather than seeing a river flowing past their homes.

Claiming treaty rights

Indian actions tied directly to their long-standing treaty rights have also stimulated concerns about the eco-system by gaining national attention. This occurred when actions by the states of Washington, Oregon, Michigan, and Wisconsin threatened their treaty guaranteed fishing rights. During the 1850s tribal leaders in each of those states had signed treaties guaranteeing them the right to fish at their traditional sites with traditional methods and equipment. A century later, alarmed by the shrinking numbers of Pacific salmon and Great Lakes trout, commercial and sport fishing groups in all four states demanded new restrictions against Indian fishing. In response, state authorities moved to limit tribal members' fishing. The new regulations struck directly at basic tribal environmental treaty rights, cultural practices, and food sources. Chiefs in all four states knew of their existing treaties, and they declared their right to fish off the reservations and beyond the state-imposed fishing seasons. As a result, Indian fishermen continued gathering salmon, lake trout, and walleye pike, despite the new state rules. Their actions prompted state officials to arrest, jail, and fine them, often seizing their boats and equipment at the same time (Harmon 218–225).

In 1964 Oregon Judge Robert Jacques issued an injunction prohibiting the Indians' net fishing in Puget Sound on the Pacific coast. Asserting their treaty rights the Nisqually tribe openly defied Washington authorities. The widespread arrests by state officers attracted national media attention when civil rights advocates including entertainment stars Marlon Brando, Jane Fonda, and Dick Gregory joined the protestors. (Heffernan 115–118; Wiklinson 169–170). The controversy spread to Oregon in 1969 when the US Circuit Court there ruled that state limits on Indian fishing ignored the Indians' special treaty rights. That decision encouraged the Washington tribes to renew their net fishing, leading to more state violence and multiple arrests. Two weeks after the decision, the US Justice Department moved to support a suit by seven tribes to protect their treaty rights. That brought an Indian victory in the 1974 *United States v. Washington* (Boldt decision). The ruling upheld native fishing rights and awarded the tribes one half of all the salmon caught in Puget Sound. Just over twenty years later, in 1995, Judge Edward Rafeedie extended the Boldt ruling by allowing Indians to harvest one half of the Puget Sound shellfish (Wilkinson 200–202; Parman 164–165). In these instances, the Indians used physical confrontations which gained wide national media attention. Having gotten that, they shifted tactics to legal action based on their environmental treaty rights and cultural practices. Their actions led to widespread attacks, including attempted bombings of their boats, and firing shots at Indian fishermen (Harmon 218–244; Parman 164–165, 176). The Puget Sound controversy kept public awareness of ecological issues high in the region for several decades.

Meanwhile, in Michigan William Jondreau, an Ojibwe tribal member, sued state game regulators for enforcing the fishing laws there. He based his case on an 1854 treaty that gave his people the right to gather lake trout and other fish wherever and whenever they wanted to (Michigan Fishing Dispute 57). In 1971 the Michigan supreme court agreed. As the Boldt decision in Washington had done, this ruling led to vigilante attacks by white fishermen who smashed Indian boats and cut their nets. State game officers added to the chaos by arresting many tribal fishermen and seizing their equipment. The dispute ended temporarily with the 1976 Michigan Supreme Court *People v LeBlanc* decision which allowed the Indians to continue fishing, but the controversy continued for nearly another decade. Later the federal government joined the tribal suit, resulting in the 1985 *United States v. the State of Michigan* decision which upheld the treaties (Doherty 60–64, 67–68).

Indian legal victories in Michigan attracted much negative publicity over ecological questions there. State game and fish officials reacted angrily to the rulings. They had spent vast time, effort, and millions of dollars during the preceding twenty years trying to reestablish the Great Lakes sport fishing industry. It had suffered an ecological disaster with the introduction of lamprey eels and alewife infestations that resulted from the 1950s opening of the St. Lawrence seaway. Local sports fishing groups joined resort owners and others in the fishing-related tourist industry in denouncing what they called special treatment for Indians. Groups such as the Grand Traverse Area Sport Fishing Association, Save Our Bay, and the Stop Gill Netting Association generated much anti-Indian publicity. While identifying themselves as peaceful organizations of concerned sportsmen, some of their members committed vigilante raids on Indian fishing camps. At one point, statements at public meetings of the Stop Gill Netting Association became so anti-Indian that the *Record Eagle* the local newspaper complained. Its editor suggested that the group's rhetoric had become so bigoted that its members "should cover their bodies with white sheets [and] get a charter from the Ku Klux Klan" (Doherty 83). This failed to calm the situation. In 1982 the public relations director for the Grand Traverse Sport Fishing Association told its members that they and the" State have exhausted all legal options that [gill] netting and violence appears to be the only recourse left" (Doherty 83).

Next door in Wisconsin Indian activism kept ecological matters in the news as well. There, citing treaties from the 1840s and 1850s, six bands of Ojibwe people began to insist on their guaranteed fishing rights. Those included being able to spear fish for walleyed pike, a highly desirable game fish. In 1973 several men from the Courte Oreilles band cut holes in the ice and began to spear the fish. State law prohibited spear fishing, so game officers arrested the men. When a local court convicted them of poaching because they had fished out of season, the band government defended its members citing their 19th century treaty. They sued and in the 1987 decision *Lac Courte Oreilles v. Wisconsin,* the court ruled that the Indians could use modern equipment and could sell their catch if they chose (Nesper 89).

As it had happened in Washington and Michigan, the ruling infuriated local sport fishermen and resort owners who claimed that Indian spearfishing endangered the fish supply and destroyed the local economy in small towns dependent on vacationing white sportsmen. They turned to vigilante actions. A local organization —Stop Treaty Abuse/Wisconsin -- used violence and intimidation, hoping to end the spearfishing. Their protests included posters that said, "Spear an Indian Save a Walleye." In response young Indian men labeled themselves "Walleye Warriors" and defended the tribal fishermen as violence continued. The dispute got so much public attention that it spilled over into politics. During his run for reelection, Wisconsin Governor Tommy Thompson used the controversy directly. He campaigned against Indian treaty rights, opposed spearfishing, and met with and encouraged leaders of vigilante anti-Indian groups. In 1991 the US District Court issued an injunction against the Stop Treaty Abuse, the most active vigilante group (Nesper 210). At the same time a state fish & game department study found that Indian spear fishing took so few fish that the practice posed no danger to the walleye population, and the controversy faded. Clearly this dispute over Indian fishing practices kept public interest and awareness of their close associations with environment matters.

Gathering public support

The repeated statements of tribal sovereignty, appeals to treaty rights, and claims of protection offered by the 1987 American Indian Religious Freedom Act, all combined to reinforce the public stereotypes about Indians as environmentalists. The cumulative effect of this became clear during what became a national uproar in 2016 over the Dakota Access Pipeline. That action began in February 2016 when the US Army Corps of Engineers approved the petition of the Texas-based Energy Transfer Partners to move its partially built oil pipeline away from Bismarck, North Dakota. The company proposed the move in response to North Dakota officials who feared that possible leaks would threaten the city's drinking water. Under the new plan the company proposed shifting its pipeline south across land near the Standing Rock Sioux Indian Reservation. There it would pass under Lake Oahe on the Missouri River and near old Indian burial grounds. The change brought negative responses from the Environmental Protective Agency, the Department of the Interior, and the President's Advisory Council on Historic Preservation immediately. Each of them demanded that the Corps of

Engineers examine alternative routes and issue a formal Environmental Impact Statement (Braine 7).

Sioux leaders objected to the new location at the same time. Seeing it as a desecration of sacred burial grounds, more importantly, as a threat to their water supply, their response followed well-established patterns of resistance. First, they turned to social media calling for backing by other Native groups and environmentalists. This attracted support from more than 300 recognized tribes as well as Indigenous groups from Latin America Australia, New Zealand, and the Sami in Finland. (Liu 9; Medina). In April 2016 Sioux elder Ladonna Brave Bull Allard, Standing Rock Historical Preservation Officer, led protestors in establishing Sacred Stone Camp and began demonstrating against the new pipeline route. Then, joining forces with Earthjustice, an environmental group, the Standing Rock Sioux Tribe filed suit against the Corps of Engineers. Their suit alleged that the pipeline violates both the National environmental Policy Act and the Clean Water Act. The court rejected the suit almost immediately. (McKenna 4).

During July 2016 construction crews began clearing burial grounds the Indians considered sacred and Sioux rushed to the site. There security guards using attack dogs charged the protestors. Witnesses filmed the incident and several million people saw the event on You Tube and other media outlets. That led other tribes to support the pipeline opponents and people from dozens of tribes hurried to join the Sacred Stone Campers. Soon nearly 4,000 Indians from other tribes joined the Sioux, making the camp one of the largest gatherings of American Indians in modern American history (Liu).

In September of that year pipeline workers using bulldozers dug up land that included a Sioux burial ground, and protesters blocked the site, halting work. At that point local police and sheriff's officers moved in to remove the demonstrators, but the protests continued. On October 27, this led to violence as North Dakota state troopers, Sheriff's deputies, and National Guardsmen resorted to teargas, pepper spray, water cannons, rubber bullets, tasers, concussion grenades, and biting dogs against protesters, while several million people viewed the attacks on YouTube (*Arizona Daily Star,* 28 Oct 2016). This heavy-handed military response to the demonstrators brought repeated condemnation of local officials. Amnesty International USA criticized the action saying that "confronting men, women, and children while outfitted in gear more suited for the battlefield is a disproportionate response" (Amnesty International USA). The UN Human Rights Council heard Standing Rock Chairman David Archambault II cite treaties that recognized tribal sovereignty, as he explained the resistance as an effort to protect tribal sovereign and religious rights. A UN official called on US authorities to "protect the right to freedom of peaceable assembly of Indigenous people" (Germanos). Despite the public outcry President Donald Trump authorized the Corps of Engineers to restart the work (Jones, Diamond and King).

Conclusion

The examples discussed here demonstrate that major Indian protests based on treaty rights had a better chance of success in the courts than those basing their dissent on other ideas. At the same time disputes with a clear environmental component attracted far more public notice and support than any others based on either treaty rights or religious sites. As the 2016–17 uproar over the Dakota Access Pipeline showed Indian protests over their water rights brought wide public interest and acceptance. Clearly the broad-scale and often sympathetic media coverage of the events in Dakota indicated the growing public interest in environmental matters. It resulted from public views of Indians as ecologically sensitive and as protectors of the environment that had developed during the preceding eighty years, as promoted by the "Crying Indian" symbol.

References

American Friends Service Committee. *Uncommon Controversy: Fishing Rights of the Muckleshoot, Puyallup and Nisqually Indians.* Seattle: University of Washington Press, 1970. Print.

"Amnesty International USA to Monitor North Dakota Pipeline Protests." http//:www.amnestyusa.org/news/press-releases/amnesty-international-usa-to-monitor-north-dakota-pipeline-protests. Web.

Arizona Daily Star 2 Dec. 2018: C3. Print.

Armas, Genaro C. "Erosion at Seneca cemetery dredge lingering bitterness." http://indiancountrynews.net/index.php?option=com_content&task=view&d=869&Itemid=116. Web.

Benally, Jeneda and Jean Goodman. "Native Americans Fight to Save Sacred Site." https://www.culturalsurvival.org/publications/culgturfal-survival-Quarterly/. Web.

Berkhofer, Robert F. *The White Man's Indian: Images of the American Indian from Columbus to the Present.* New York: Vintage Books, 1978. Print.

Braine, Theresa. "Three Federal Agencies Side with Standing Rock." *Indian Country Today* (April 27, 2016): 7. Web.

Cash, Johnny. "Bitter Tears: Ballads of The American Indian." New York: Columbia Records, 1964.

Doherty, Robert. *Disputed Waters: Native Americans & the Great Lakes Fishery.* Lexington: University of Kentucky Press, 1990. Print.

Finnerty, Megan. "Compromise complicated on debate over faith, water, land." https://The Republic azcentreal.com. March 13, 2015. 6–7. Web.

Germanos, Andrea. "UN Experts to United States: Stop DAPL Now." http://www.commondreams.porg/news/2016/09/25?un-experts-united-states-stop-dapl-now. Web.

Gray, Jim. "Standing Rock: The Biggest Story That No One's Covering." *Indian Country Today.* 8 Sept 2016. Web.

Harmon, Alexandra. *Indians in the Making: Ethnic Relations and Indian Identities around Puget Sound.* Berkeley: University of California Press, 1998. Print.

Hauptmann, Laurence H. *In the Shadow of Kinzua: The Seneca Nation of Indians since World War II.* Syracuse: Syracuse University Press, 2014. Print.

Heffernan, Tova. *Where the Salmon Run: The Life and Legacy of Billy Franks, Jr.* Seattle: University of Washington Press, 2012. Print.

Helfrich, Joel, Dwight Metzger, and Michael Nixon. "Native Tribes Struggle to Reclaim Sacred Sites." *Twin Cities Newspaper,* June 1, 2005. http://w.w.w.mountgraham.org/content/star-struck-astronomical-abuse-indigenous-sacred-sites. Web.

Hoover, William E. *Kinzua: From Cornplanter to the Corps.* Lincoln: University of Nebraska Press, 2005. Print.

Johansen, Bruce E. *Enduring Legacies: Native American Treaties and Contemporary Controversaries.* Westport, CT: Praeger, 2004. Print.

Jones, Athena, Jeremy Diamond, and Gregory Krieg. "Trump advances controversial oil pipelines with executive actions." *CNN,* January 24, 2017. http://cnn.com/2017/01/24/politics/trump-keystone-xi-dakota-access-pipelines-executive-actions/index.html. Web.

Josephy, Alvin M. Jr. "Cornplanter, Can You Swim?" *American Heritage.* 20.1 (1968): 1–7. Web.

Kelley, Alexandra. "Despite campaign promise, Biden administration will not shut Down Dakota Access Pipeline 'at this time.'" Changing America. April 12, 2021. https//the hill.com/changing-america/sustainability/environment 5347730-despite-campaign-prfomise-biden-administration-will-not-shut-down-Dakota-Access-Pipeline-at-this-time. Web.

Kennedy, John F. "Letter to the President/ of the Seneca Nation of Indians Concerning the Kinzua Dam on the Alleghney River." August 9. 1961. http//:www.presidency.ucsb.edu/ws/index.php?pid=8279. Web.

Kretch, Shepard III. *The Ecological Indian: Myth and History*. New York: W. W. Norton and Company, 1999. Print.

Lawson, Michael L. *Dammed Indians: The Pick-Sloan Plan and the Missouri River Indians, 10944-1980*. Norman: University of Oklahoma Press, 1982. Print.

Liu, Louise. "Thousands of protestors are gathering in North Dakota—and it could led to 'nationwide reform.'" *Business Insider,* September 13, 2016. http//:www.businessinsider.com/photos-north-dakota-pipeline-protest-2016-9. Web.

McKenna, Phil. "Native American Pipeline Protest Halts Construction in N. Dakota." *Inside Climate News* 19 Aug. 2015. 4. Web.

MacPherson, James and Blake Nicholson. "Police evict North Dakota pipeline Protestors." *Arizona Daily Star* [Tucson] 28 October 2016 A13. Print.

Medina, Daniel A. "Dakota Pipeline Company Buys Ranch Near Sioux Protest Site." NBC NEWS, 24 Sept. 2016. http://www.nbcnews.com/news/us-nrews/dakota-pipeline-company-buys-ranch-near-siouxprotest-site-records-n65051. Web.

"Michigan Indians in Fishing Dispute." *New York Times*. 3 Oct 1971. 57. Web.

Mount Graham: Science and Apache Religion. http://nativeamericannetroots.net/diary/471. P. 2. April 17, 2010. Web.

"Navajo Nation v. United States Forest Service, 535 F.3d 1058 (9th Cir. 2008). Web.

Nesper, Larry. *The Walleye War: The Struggle for Ojibwe Spearfishing and Treaty Rights*. Lincoln: University of Nebraska Press, 2002. Print.

Park, Madison and Mayra Cuevas. "Dakota Access: Pipeline clashes turn violent." http://www.cnmnm.copm/2016/11/21/us/dakota-access-pipeline-protests/index.html. Web.

Parman, Donald L. *Indians and the American West in the Twentieth Century*. Bloomington: Indiana University Press, 1994. Print.

Pearce, Roy Harvey. *Savagism and Civilization: A Study of the Indian and the American Mind*. Berkeley: University of California Press, 1953. Print.

Protect the Peaks. http// protectthepeaks.org/facts/12/18/2019. Web.

Sage, L. "A Brief History of the Controversy Surrounding the Mount Graham International Observatory." In A. Heck, ed. *Organizations and Strategies in Astronomy,* 4 (2003): 75–91. Dordrecht, Netherlands: Kluwer Academic Publishers, 2003. Print.

Shanks, Bernard D. "The American Indian and Missouri River Developments 1," *Journal of the American Water Resources Association* 10.3 (1974): 573–579. Web.

Smith, Mary H. "Wilson v. Block." *Natural Resources Journal*, 26. 1 (1986): 169. Web.

Sainte-Marie. *The Buffy Sainte-Marie Songbook.* New York: Grosset & Dunlap. 1971. Print.

Stromberg, Juliet and Duncan T. Patten, "Vegetation Dynamics of the Spruce-Fir Forests of the Pinaleño Mountains." 89–99 in Istock, Conrad A. and Robert S. Hoffmann, eds. *Storm over a Mountain Island: Conservation Biology and the Mt. Graham Affair.* Tucson: University of Arizona Press, 1995. Print.

Swetnam, T. W. and P. M. Brown. "Oldest known conifers in the Southwestern United States." In *Old-Growth Forests in the\ Southwest and Rocky Mountain Region.* US Forest Service General Technical Report RM 213: 24–38 (1902). Web.

Toward Castle Films. "Lake of Betrayal." http://towardcastlefilms.com/filmography/lake-of-betrayal. 2017. Web.

U. S. Forest Service. *Final Environmental Impact Statement for Arizona Snowbowl Facilities Improvements,* Vol. 1. Web.

Wilkinson, Charles. *Blood Struggle: The Rise of Modern Indian Nations.* New York: W. W. Norton, 2005. Print.

Loredana Bercuci

Matter, Objects and Nature Knowledge in William Byrd II's *Histories of the Dividing Line betwixt Virginia and North Carolina*

Abstract: William Byrd II's *Histories of the Dividing Line betwixt Virginia and North Carolina* (1728/1929) is an account of the surveying of the border between Virginia and North Carolina that attempted to settle the dividing-line dispute between the two colonies. In this study, I discuss the intersections between environmental knowledge, race, and objects in William Byrd II's *Histories* from the perspective of material ecocriticism in order to show how the objectification of the boundary line works to objectify non-white races, as well as nature, and to conceal whiteness in colonial America.

Introduction

William Byrd II's *Histories of the Dividing Line betwixt Virginia and North Carolina* (1728/1929) is an account of the surveying of the border between Virginia and North Carolina that attempted to settle the dividing-line dispute between the two colonies. Chartered in 1606 and settled in 1607 by the Virginia Company, the Colony of Virginia was the first successful English colony in North America, which garnered it the name of "Old Dominion." It functioned as a crown colony, i.e. a colony administered by the monarch who appointed a colonial governor. The Province of North Carolina, on the other hand, was a proprietary colony chartered in 1663, that is, the type of colony in which the monarch, in this case Charles II, awarded land to a proprietor, who was in effect the ruler of the territory in question. Having been established much later than Virginia, North Carolina's boundary was set in relation to that of its Southern and Northern neighbors, Spanish Florida and Virginia. Before the charter was issued for North Carolina, some Virginians had already settled the land southwards, beyond the line that would demarcate the border between the two colonies. In 1665, another charter was issued for North Carolina, which moved the boundary 30 miles to the North, further escalating the border dispute.

In the introduction to the 1929 edition of Byrd's account, William K. Boyd notes that the legal conflict depicted in Byrd's *Histories* began in 1680, "when certain people on the border lands refused any longer to pay the Virginia quit rents, although their titles were from the Virginia land office" (xvii). For North Carolina, it was not only taxes to be received from these inhabitants that were at stake, but also access to the Nottoway River, which would aid tobacco trade – a vital commodity for the newer colony's economic success (Boyd xix). For the next couple of decades, Virginia did not recognize the second charter and, consequently, did not take any action to establish an official border through a physical line until 1705 (Boyd xviii). A first joint commission of the two colonies was appointed to survey the land and establish the line in 1710, when the Lords of Trade and Plantations recommended it. This commission failed to produce results both colonies would approve. A second commission, of which Byrd was part, along with William Dandridge, and Richard Fitz-William, as representatives of Virginia, was appointed in 1728; on the side of North Carolina, Christopher Gale, Edward Moseley, William Little, and John Lovick were appointed. The commission first met at Currituck Inlet, on 5 March 1728. The line was carried by the joint commission, which surveyed and marked the land, in two sessions (in March and in September) to the Roanoke River. It was carried further into the Appalachians to Peter's Creek by the Virginia commissioners alone.

At the heart of this dispute was Weyanoke Creek, the final point named in the charter of 1665 from which the border was to extend to the West (Boyd xvii). The original name of the creek had been changed, so that it could not be located. As such, the Virginians identified it as the Wicocon Creek, while the North Carolinians argued that it was the Nottoway River, each attempting to secure more of the disputed land for themselves. Before a joint commission was appointed, Virginia attempted to conduct a secret survey "to secure the affidavits of old residents regarding the location of Weyanoke Creek" (Boyd xviii). According to Angela Calcaterra, the members of the 1710 commission examined "Tributary Indians about the location of Weynoke Creek" (238). The river had been named after the Weyanokes who had at one point inhabited that land. Citing Robinson's *Virginia Treaties, 1607–1722* (1983), Calcaterra shows that the recorded testimony of the Weyanoke Indians interviewed "gives details of their people's place-based experiences and references the boundary dispute only obliquely" (239). This suggests an intersection of discourses and types of historiography, and that the colonists established their authority by appropriating Native environmental knowledge in the form of both testimonies and maps. Putting forth a type of narrative map by describing the trip of the commission into the Appalachians, Byrd's account, too, borrows from Native environmental knowledge.

Furthermore, as the line is carried forward, it intersects trails and borders created by different cultures, which it disrupts. During this process, the line, in fact, disrupts the material world, i.e. the environment (fields, plantations, other routes, trails, rivers, swamps), turning them into things through objectification, as the line itself becomes an object. At the same time, as it crosses an unstable, in-between territory, the boundary objectifies race (whites and non-whites alike), foregrounding the instability of object formation. Similarly, William Byrd II's *Histories of the Dividing Line betwixt Virginia and North Carolina* (1728/ 1929) may be read as an object of memory containing autobiographical elements. In this essay, I discuss the intersections between environmental knowledge, race, and objects in William Byrd II's *Histories* from the perspective of material ecocriticism in order to show how the objectification of the boundary line works to objectify non-white races, as well as nature, and to conceal whiteness in colonial America.

Material ecocriticism, objects, and whiteness

Byrd II's autobiographical text dates back to the 18th century and reflects a modern conception of the self and of nature. The *Westover Manuscripts*, published in 1841, include, in addition to Byrd's *History*, the author's other autobiographical writings. The main purpose of these works appears to be the creation of Byrd's persona as that of a man of exceptional importance and stature. When Byrd failed to achieve nobility in English society, he focused his ambitions on Virginia, aiming to succeed in politics: he attempted to buy the governorship of the colony (Kaeuper 121 122). The image crafted by his writing would have been used to aid his pursuits of power. Moreover, Byrd's constant categorization in his works, both of humans and of the natural world, betrays a type of logic typical of modernity as "a total separation between nature and culture" (Latour 30).

Byrd's text reflects modern conceptions of human progress in which the material world necessitates "taming" or is, at most, a backdrop for human agency. To read his text, I employ the concept of the Anthropocene as the

> manifest of the dark underside of 'modernity', its destruction of the environment, its deep implication in imperialism, colonialism and neo- colonialism, as well as in modes of hierarchy and discrimination that condemn billions to poverty. (Clark 24)

Therein, "nature" "functions deceptively as the essentially political notion of a condition supposedly prior to human politics" (Clark 32). Additionally, "nature" is correlated with "wilderness" and pristine state from which indigenous inhabitants are conspicuously absent. In other words, "the Enlightenment view of nature is inextricably tied to colonial European ambitions to dominate the world" (Hornborg 96).

In exploring the natural world in connection to notions of race, I draw on material ecocriticism, which sees "the world's material phenomena as knots in a vast network of agencies, which can be 'read' and interpreted as forming narratives, stories" (Iovino and Oppermann 1). Matter is endowed with agency as a "material 'mesh' of meanings, properties, and processes, in which human and non-human players are interlocked in networks that produce undeniable signifying forces" (Iovino and Oppermann 1). In Byrd's rendering, I argue, both nature and racial others are turned into objects.

According to Brown (2001, 2016), there is a distinction between objects and things. In essence, the existence of an object is dependent on the presence of a subject. Objects "are codes by which our interpretive attention makes them meaningful, because there is a discourse of objectivity that allows us to use them as facts" (Brown "Thing Theory" 4). Furthermore, the otherness of the object and its stability "stabilizes the objecthood of the self" (Brown *Other Things* 21). Things, on the other hand, are "some underorganized material field or some unorganized amalgam or mass: a field of sensations before they are organized into discrete objects" (Brown 2016: 22). An object becomes a thing when it loses its function as an object. The main effect of things is to illuminate the relationship between the subject and the object: "The story of objects asserting themselves as things, then, is the story of a changed relation to the human subject and thus the story of how the thing really names less an object than a particular subject-object relation" (Brown 2001: 4). In the complex relationship between subject and object, the object formation process may reveal strategies of hierarchization on different levels.

In terms of race, the objectification process is related to othering. Whiteness, consequently, defines the subject in opposition to non-whites, who become objects in the process. In American history, as James Baldwin puts it, "America became white - the people who, as they claim, 'settled' the country became white - because of the necessity of denying the Black presence, and justifying the Black subjugation" (178). This process of othering is also taken up by Toni Morrison when she asserts, in a 1997 essay, that in the "construction of blackness and enslavement could be found not only the not-free but also, with the dramatic polarity created by skin color, the projection of the not-me" (82), suggesting that the white subject was born in dialectical relationship with the non-white presence. In other words, as Sherrow Pinder has phrased this idea, "[w]hiteness is a dialectical force, which requires the nonwhite presence in order to maintain its malignant existence" (7).

Paradoxically, in this process of othering, both whiteness and non-whiteness need to be defined by invisibility. The invisibility of whiteness, on the one hand, requires whiteness not to be seen by the non-whites, i.e. their own invisibility. Referring to the white master – black slave relationship, bell hooks states that "[t]o look directly was an assertion of subjectivity, equality. Safety resided in the pretense of invisibility" (340). Richard Dyer similarly argues that the "white discourse implacably reduces the non-white subject to being a function of the white subject, not allowing her/him space or autonomy" (13). As such, non-whites are robbed of subjectivity (they are objectified), so that whiteness can both exist and become invisible and, consequently, non-whites could be owned as objects. The invisibility of whiteness allows it to be indistinguishable from subjectivity. The invisibility of whiteness gives the illusion of "infinite variety" (Dyer 13) of subjectivity. Additionally, invisibility allows whiteness to be normative on grounds of universality: "as a way of seeing and knowing the world that masquerades as universality and remains largely unnamed and unrecognized" (Watson 5).

In what follows, I will discuss how these elements intersect in William Byrd II's *Histories of the Dividing Line betwixt Virginia and North Carolina* (1728/ 1929) suggesting that the dividing line as a thing pertaining to the material world disrupts other objects, turning them into things, because it becomes an object itself. Mirroring that process, whiteness turns non-whites into objects while it maintains its subject position.

William Byrd II's autobiographical rendering of a narrative map

William Byrd II's *Histories of the Dividing Line betwixt Virginia and North Carolina* (1728/ 1929) is, as stated above, an autobiographical account, a narrative cartography of border drawing and dispute between Virginia and North Carolina. As such, the reader would take it as truthful, if certain contradictions engendered by the 1929 edition (and almost all subsequent editions, which have typically kept its form) are disregarded. The 1929 edition contains the *History of the Dividing Line* alongside *The Secret History of the Line,* which contains Byrd's diary entries that narrate the same events, sometimes in more, sometimes in less detail. William K. Boyd, the editor of the 1929 edition and the first to publish the works side by side, meaning the *History* on the left side of the page and the *Secret History* on the right side, argues that *The Secret History*, considering its reduced length, was most likely written first as notes for the extended version which Byrd wanted to publish in book form (Boyd xv).

There are several discrepancies between the two versions. The most obvious difference is that in the *Secret History* Byrd uses code names for all the characters that appear in the story. Another discrepancy relates to two members of the commission (Richard Fitz-William, the Surveyor-General of Customs for the Southern Colonies, and Alexander Irvine, a Mathematics professor charged with surveying measurements), whom Byrd disliked and distrusted and whom he took considerable liberties in portraying negatively in the *Secret History*, which includes the scene of a brawl between the commissioners having to do with recognition of authority. The *Secret History* also contains a number of documents, such as letters, not included in the *History*. Most importantly, the *Secret History* includes scenes of violence towards women living on the frontier, including rape. The *History*, on the other hand, contains considerable additional information about the flora and fauna of the territory surveyed, about the customs of the indigenous peoples encountered along the way, and about the lives of the various other inhabitants of the borderlands.

This doubling of the history lines makes the account appear less, yet, at the same time, more reliable. While the *Secret History* reveals some unreliable elements included in the *History*, their inclusion alongside it gives the impression that it conceals nothing. For example, the *History* depicts Edenton, a town that does not appear in the *Secret History* and that Byrd never visited, in negative colors. As readers, we doubt the veracity of the town's description. Consequently, the reader questions the veracity of the *History*, but sees the *Secret History* as the real account due to the fact that, being a diary, the *Secret History*, though highly subjective, could be more authentic. In reality, there is no reason to consider the *History* the less reliable version, or to question both accounts. As it stands, the autobiographical narrative is strengthened and, while the *History* to which our attention is drawn becomes the product of memory, *The Secret History* remains its object.

In a mirroring process, the dividing line between the two colonies is an imaginary border objectified into a physical map. By being measured and drawn, the line is endowed with material existence: It becomes a thing, besides being a symbol or a convention by means of which the two colonies are separated. From the very beginning of the account, the two parties cannot agree on the physical place where the line should start or on its exact coordinates. Virginia wanted to start the line at the Spit of Sand on the North Shore of Currituck Inlet, whereas North Carolina insisted on starting it 200 yards beyond. One party found the latitude to be 36° 31' while the other put it at 36° 30'. Further drawing attention to its thingness, the line is physically drawn, and this becomes especially apparent when it fades: "The line was marked only on trees and in time disappeared or became vague" (Boyd xxiii). As such, from the very beginning of the narrative, the material world opposes the colonial action of surveying and categorizing, which foregrounds the process as one in which both human and non-human actors are involved. Subsequent descriptions of the material world are similar, especially as they come into contact with the line.

Attention is further drawn to the thingness of the line when material objects resist it throughout the narrative. As the men advance, they face opposition from the natural world in the form of rivers, thickets, and swamps. This is particularly obvious when they reach what is named the Dismal Swamp, which is extensively and hyperbolically described. Consider the following excerpt:

> The Skirts of it were thinly Planted with Dwarf Reeds and Gall-Bushes, but when we got into the itself, we found the Reeds grew there much taller and closer, and, to mend the matter was so interlac'd with bamo-briars, that there was no scuffling thro' them without the help of Pioneers. At the same time, we found the Ground moist and trembling under our feet like a Quagmire, insomuch that it was an easy Matter to run a Ten-Foot-Pole up to the Head in it, without exerting any uncommon Strength to do it. (Byrd II 62)

Interestingly, it is the instability of the ground that poses the greatest resistance, giving the impression of ungraspable materiality. Moreover, it is mentioned that the Pioneers, who had advanced the colonies into the borderlands, are best suited to wrestle with its stubborn thingness. This foreshadows how these very same pioneers will be portrayed later on, when they are defined by an unstable (racial) identity.

Other elements whose thingness is exposed by the advancing line are the plantations that stand in its way. In the process of drawing the border between the colonies, the plantations are divided, with one part remaining in one colony and the other in the neighboring one, thus being accountable either to both legislatures or to neither. As it is evident from the following excerpt, the border mapping placed the borderland inhabitants in an unstable situation, since, having their property divided by the line they actually belonged to both colonies or to neither:

> The Line cut William Spight's Plantation in two, leaving little more than his dwelling House and Orchard in <u>Virgina.</u> Sundry other Plantations were Split in the same unlucky Manner, which made the Owners accountable to both Governments. Wherever we passed we constantly found the Borderers laid it to Heart if their Land was taken into Virginia: They chose much rather to belong to Carolina, where they pay no Tribute, either to God or to Caesar. (Byrd II 104)

Due to the disruptions caused by the dividing line, the borderland inhabitants could no longer consider themselves Virginians or North Carolinians. They were both or neither. In either case, they were objectified, they became "things."

As it advances along the line, the commission encounters the material marks of the othered, the Native Americans, whose visibility matches the other things met along the line. In the following excerpt, Byrd describes the trading path the commission comes across in their journey:

> We crost the Indian Trading path above-mention'd about a Mile from our Camp, and a Mile beyond that forded Haw-Tree-Creek. The Woods we passed thro' had all the Tokens of Sterility, except a small Poison'd Field, on which grew no Tree bigger than a Slender Sapling. (Byrd II 160)

Such instances are quite frequent in the account, revealing how the other, whether objectified nature or beings could be destroyed.

In another scene, a Native war practice materially hinders the progress of the surveyors: "[t]he Smoak continued still to Veil the Mountains from our Sight, which made us long for Rain, or a brisk Gale of Wind, to disperse it" (Byrd II 224). As Byrd explains, the smoke had been caused by "the fireing of the Woods by the Indians, for we were now near the Route the Northern Savages take when they go out to War against the Cataubas and other Southern Nations" (224). The denigration of the Natives' practices fits into the colonizers' portrayal of the other as uncivilized in the attempt to foreground the "correct" and "civilized" practices of the colonists.

In spite of this, the members of the party advance with the help of a Native guide, and are physically supported by his knowledge of the land and hunting – there are several instances when the commissioners would have faced starvation, but for the skills of the Native who hunts and cooks for them. The guide is called "our Indian" throughout the narrative, suggesting an objectification process through which he was transformed into property. At the same time, Bear-skin's (the name by which they address the guide) skills are frequently praised as Byrd argues: "We cou'd not entirely rely on the Dexterity of our own Men" (160). His nature knowledge shows him as both a noble savage and as animalistic. He is said to be in a "State of Nature, without one Glimpse of Revelation or Philosophy" (202), and is described as "Gross and Sensual" (202).

Akin to the advancing line, the surveyors themselves, as representatives of whiteness, expose the racial instability encountered along the way. As the line turns into an object, so do those who are racially othered. Two scenes from the account are particularly revealing for this, quoted at length below. The first scene discloses that escaped slaves had the habit of settling on the borderlands:

> There we came upon a Family of Mulattoes, that call'd themselvs free, tho' by the Shyness of the Master of the House, who took care to keep least in Sight, their Freedom seem'd a little Doubtful. It is certain many Slaves Shelter themselves in this Obscure Part of the World, nor will any of their righteous Neighbours discover them. On the Contrary, they find their Account in Settling such Fugitives on some out-of-the-way-corner of their Land, to raise Stocks for a mean and inconsiderable Share, well knowing their Condition makes it necessary for them to Submit to any Terms. (Byrd II 56)

Upon meeting black people, Byrd immediately assumes that they are slaves, i.e. property, othered objects. He draws attention to their physicality, marking them racially by their skin color. Furthermore, in line with bell hooks' account of the invisibility of slaves, Byrd discursively attempts to render them invisible, and thus dismantle their subjectivity as he refers to their "Shyness." Additionally, his account of what happens to fugitives places black bodies in a state of perpetual ownership: even escaped slaves continue to be property.

In the second scene, describing white inhabitants on the borderlands, Byrd others them as non-whites. As Matt Wray suggests, "they were considered different categories, but between the lower classes and the lower races, there was considerable overlap in the symbolic properties, characteristics, and traits ascribed to each" (22–23). Note the excerpt below:

> It approaches nearer to the Description of Lubberland than any other, by the great felicity of the Climate, the easiness of raising Provisions, and the Slothfulness of the People. Indian Corn is of so great increase, that a little Pains will Subsist a very large Family with Bread, and then they may have meat without any pains at all, by the Help of the Low Grounds, and the great Variety of Mast that grows on the High-land. The Men, for their Parts, just like the Indians, impose all the Work upon the poor Women. [...] When the weather is mild, they stand leaning with both their arms upon the corn-field fence, and gravely consider whether they had best go and take a Small Heat at the Hough: but generally find reasons to put it off till another time. (Byrd II 92)

He describes them as lazy, "just like the Indians," and attributes this character trait to their fortunate situation, their inhabiting a fertile land. Following the logic of scientific racism, Byrd seems to suggest that it is the climate that awards certain traits to people. In addition, the subject and the object seem to switch places: the material world becomes the subject that does the work, while the humans are passive. In this way, land, which can be owned, is aligned with whiteness, while the borderline inhabitants, who do not conform to the status quo, who are engaged in boundary crossing both racially and materially, are racialized and thus turned into objects.

Conclusion

William Byrd II's *Histories of the Dividing Line betwixt Virginia and North Carolina* (1728/1929), the story of the cartography of the line dividing Virginia and North Carolina, is an autobiographical account and, as such, an object of memory. The juxtaposition of the two histories objectifies one and authenticates the other. This dialectical process is mirrored in the clash between the line and the elements it encounters. As the land is surveyed and the line is carried forward, it disrupts the objects it encounters in its way, thus turning them into things. At the same time, the line turns from a thing into an object. Additionally, passing through an unstable, in-between territory, the boundary objectifies race. The complex intersections between the material world, nature knowledge, race, and objects in William Byrd II's *Histories* work both to create and to conceal anthropocentric hierarchies in colonial America.

References

Baldwin, James. "On Being 'White' … and Other Lies." *Black on White*. Ed. David Roediger. New York: Schoken Books, 1999. 177–180. Print.

Boyd, William K. "Introduction." *Histories of the Dividing Line betwixt Virginia and North Carolina*. Ed. William K. Boyd. Raleigh: The North Carolina Historical Commission, 1929. xi–xvii. Print.

Brown, Bill. "Thing Theory." *Critical Inquiry* 28 (2001): 1–22. Print.

---. *Other Things*. Chicago: The University of Chicago Press, 2016. Print.

Byrd II, William. *Histories of the Dividing Line betwixt Virginia and North Carolina*. Raleigh: The North Carolina Historical Commission, 1929. Print.

Calcaterra, Angela. "Locating American Indians along William Byrd II's Dividing Line." *Early American Literature* 46 (2011): 233–261. Print.

Clark, Timothy. *The Value of Ecocriticism*. Cambridge: Cambridge University Press, 2019. Print.

Dyer, Richard. *White: Essays on Race and Culture*. New York: Routledge, 2013. Print.

hooks, bell. "Representing Whiteness in the Black Imagination." *Cultural Studies*. Ed. Lawrence Grossberg et al. New York: Routledge, 1992. 338–342. Print.

Hornborg, Alf. "Artifacts Have Consequences, Not Agency: Toward a Critical Theory of Global Environmental History," *European Journal of Social Theory* 20 (2017): 95–110. Print.

Iovino, Serenella and Serpil Oppermann. *Material Ecocriticism*. Indianapolis: Indiana University Press, 2014. Print.

Kaeuper, Geoffrey. "New Dominance in the Old Dominion: Steadying William Byrd in "The Secret History of the Line." *The Southern Literary Journal* 36 (2003): 121–139. Print.

Latour, Bruno. *We Have Never Been Modern*. Cambridge, MA: Harvard University Press, 1993. Print.

Morrison, Toni. "Playing in the Dark: Whiteness and the Literary Imagination." *Critical White Studies: Looking Behind the Mirror*. Eds. Richard Delgado and Jean Stefancic. Philadelphia: Temple University Press, 1997. 79–84. Print.

Pinder, Sherrow O. *Whiteness and Racialized Ethnic Groups in the United States: The Politics of Remembering*. Lanham: Lexington Books, 2011. Print.

Watson, Veronica T. *The Souls of White Folk: African American Writers Theorize Whiteness*. Jackson: University Press of Mississippi, 2013. Print.

Wray, Matt. *Not Quite White: White Trash and the Boundaries of Whiteness*. Durham: Duke University Press, 2006. Print.

Dragoș Osoianu

Material Ecotheological Implications of Food Eating

Abstract: This paper explores the material ecotheological implications of food eating, especially concerning the Christian cultural system of beliefs. The theoretical background relates to Material Ecocriticism and Ecotheology, the latter integrating both a panentheistic and a pantheistic vision of post-systematic theology.

Introduction

The incipient premise of the research issues from the Biblical command of Man ruling and naming the kingdom of animals stems from his creation in God's image. This patriarchal and anthropocentric paradigm establishes a hierarchical distinction of Human and Animal, consumer and the consumed, us and them, culture and wilderness. In this sense, the human being has the means to master, produce and consume the world – animals, vegetables, minerals and water. Nevertheless, from an ecocritical and material perspective, the dichotomy between the human realm and the natural one needs to be transcended, in order to (re)discover their connections. The act of eating is not necessarily an act of enslaving the natural object by the human subject, but an act of appropriating the divine logoi. Nature has to be seen as a pervasive narrative body, in which the boundaries of inside and outside, identity and alterity are broken. All parts of the natural continuum, animals, plants or even bacteria, have a variable degree of agency. By consuming them, an agentive, material and consubstantial connection is established among them, with humans as part of this natural continuum.

In general, food is considered to be the material substance consumed to provide nutritional matter for an organism. It may consist of animals, plants/vegetables or fungi which can provide various nutrients to the material bodies, such as proteins, fats, carbohydrates, minerals and vitamins. These material substances are assimilated by the body through a complex process of ingestion and digestion. The goal of the organic act(s) of impropriating food is to release the energy needed to function, to provide the material for future growth and, in general, to maintain the miraculous processes of life. Of course, food exists in order to be eaten and, in this respect, one may find it absurd that someone should speak about food eating. Nevertheless, the act of eating relates to various biological, ecological, ethical, cultural or theological aspects of human existence and it involves more than the mechanical and inertial processes of ingestion and digestion. There is a complex metabolic circulation within and between various and different media of life, perception and consumption (Swyngedouw in ed. Heynen et al. 32).

Material ecocritical considerations

In nature, all beings are connected in terms of the medium they live in or their specific place in the food chain, as the eaten is consumed by different eaters. All places where the eater encounters the object of eating belong to Nature and, in this sense, the process of eating is ecological. There is a mutual interdependence of the eater and the eaten, because it is a matter of survival and being integrated in the larger cybernetic system of being. Every environmental subject depends on other subjects and the degree of exploitation concerning the dynamics of consumption relates to how the higher subjectivities perceive the lesser subjectivities such as animals, microorganisms or bacteria.

Thus, the ecocritical perspective is reached, one which tries to bring together literature and the environment, not necessarily as binary oppositions, but as fields of research with close connections. Since the previous century, the ecocritical theory and its pragmatic applications have undergone multiple changes. From the initial reductionist and (post)Romantic opposition of Culture and Nature, Ecocriticism has evolved into a vision in which the former is no longer divorced from the latter. The postmodern trends of the second, third and finally the fourth waves have led the theory to deconstruct the previous binary oppositions; natural wilderness is not the opposite of cultural humanity. The cultural human being is viewed as not being segregated from the natural environment, animal, vegetal or mineral, and food consumption is part of the story.

As previously implied, the final process of de-binarization is reached in the fourth wave of Ecocriticism, wherein the theory meets New Materialism and Nature is understood both as matter and material object. Prior to this material turn, the reality we cope with was thought to be constructed through language and symbols, because the object of thinking was gravitating around the human subject. On the contrary, the material ecocritical approach does not encourage a specific center, but a reflexive plurality of subjective and objective centers. In order to both exemplify and overcome the myth of Adam naming animals, the classical and anthropocentric superiority of the mental ecology over the environmental ecology needs to be reinterpreted. That is to say the human subject of thinking subsists within the same plane of existence with the object of thought – Nature, meaning animals, plants, minerals, as well as the processes of food consumption.

The pure immanence represents an inherent premise for the human subject not to be essentially different from the natural object. The cultural perspective of Matter is unstable and, thus, relates to a postmodern ethics in which there are no substantial and ethical hierarchies between the subject and object, what/who we are and what we know, ontology and epistemology (Osoianu 40). The rhizomatic non-heterogeneous ecologies of mind, culture and the environment are linked within the larger cybernetic system of Nature. The embodied human being is made from the same matter as the environmental outside, involving a symmetrical and recursive circulation between the observing subject, the observed object, what humans hunt and what humans eat as part of the overall processes. The dualism culture-nature is abolished; therefore, the hybrid text of the cultural subject is embedded into the observed fabric of reality or nature. The text as we know it becomes both cultural and natural, forming an intrinsic informational unit. At a quantic and biological level, the food we are eating is the same with us, the eaters.

The ecological materialism uses various deconstructions of opposites such as culture-nature, discourse-reality, textual-real, interior-exterior, food producers – the produced, consumers – the consumed food. In this sense, the natural outside is reconsidered as (re)animated and (re)enchanted. As seen before, the generic agency may be considered socially constructed; yet, the materialist approach considers matter not just a cultural construct, but a non-passive, generative realm of factual discourse. This generative domain of the environment resides in a continuous materialization and becoming, through production of agency and/or autopoiesis. The animals we hunt, the plants we cultivate and the food we eat also exert agency over the human subject.

The agency of matter represents an interplay of the human kingdom and the nonhuman one. The vivid dynamics of the cultural discourse and of the natural matter relates to a production of its own environmental logoi and story. There are two parts of this narrative: the first text relates to the human/subjective perception of nature whereas the second one relates to the agentive new significations which enter the human-cultural text from the environmental object. The ontological subject and the epistemological object interact and the logic of this interaction consists in forming a common ground of mutual textual meanings. Thus, nature itself becomes or, at least, can be understood as a corporeal text, full of new meanings which enter and change the human text. This ontological continuum consists of non-heterogeneous texts such us and them, our body and the food needed to construct the human bodies, our logos of understanding the world and the logoi from this world we live in:

> There are also geological, biological, and cosmic stories that compel us to envision the physical world as storied matter teeming with countless narrative agencies that infiltrate every imaginable space and make the world intelligible. [...] By conflating our interpretive practices with the horizons of numerous narrative agencies, material ecocriticism seeks to analyze their meanings disseminated across this storied world, across the stories of material flows, substances and forces that form a web of entangled relations with the human reality. This fusion of horizons has a liberating effect of moving the human vision from the language of otherness to that of differential co-emergence. (Oppermann 57, 67).

The "storied matter" displays an agentive power, meaning that it embodies texts into the cultural narrative of the human being. The two texts are and act as following: the human text has natural origins and is constructed through cultural and social means; the environmental text is non-human and, nevertheless, forms rhizomatic assemblages with the first one. This intertextual discourse has generative powers and exerts agency in a reflexive mode. The so-called new (in fact, it is the same and the change relates to the new perception of reality) discourse is part of our (us and them) material selves. The interplay of human and environmental agencies actually represents an existential discourse of knowing and being, of perceiving and being localized in an "agentive" space (Osoianu 42). The space has no limit from an ontic and epistemic perspective, whether thinking at knowing the object of eating or being this object after consumption.

Thus, the human subject and the environmental object become co-actants of the same material discourse, sharing the same textual substance and perceiving the same material texture. Virtually, the environment is, at the same time, natural, cultural and material. Conversely, the cultural subject exists in a homogeneous (de facto) material environment, full of both human and non-human subjectivities. The organic movement between the cultural and natural domains of reality is trans-corporeal in terms of the interconnectedness of the human and non-human bodies (Oppermann 85). Through the act of eating, food becomes corporeal and brings substance to our corporeal selves, in this sense, exerting agency over us and becoming part of us.

Material ecotheological considerations – An imagistic perspective

From Material Ecocriticism, we continue the inquiry with Ecotheology, which is a form of progressive theology that pays attentions to the relationship between religion and nature, with a specific regard to postmodern environmental concerns. The main premise consists in acknowledging the causality between the decadence of the religious human being and the degradation, preservation and possible restoration of the environment. Two main theological perspectives are worth considering in this context: Christian and Hindu-Buddhist views. The former is heavily enforced through hierarchies and institutions of power, even though the Byzantine system of beliefs displays a quasi-friendly attitude towards nature. In this sense, the dogmatic vision is panentheistic in terms of a difference between the divine essence, which cannot be known or experienced and absolutely transcends creation, and divine energies, which can be experienced and touched by creatures and creation.

A further difference can be made between the divine homogeneous energies, which are transcendent and consubstantial to divinity, and heterogeneous energies, which are not consubstantial to the divine essence. In other words, Logos is the transcendent prosoponic form of the divine essence, the logoi are the transcendent, homogeneous and consubstantial manifestations of the divine essence and the material world is the heterogeneous image of these logoi. Thus, the Christian God is both transcendent and immanent through its/his logoi which are immanent (in manifestation or realization) to the materiality of Nature (Yannaras 59–61). Although transcendent, the Christian God becomes corporeal within the materiality of the world and through the incarnation of Christ. Furthermore, in terms of liturgical celebration, Christ becomes corporeal through the agency of food – wine and bread.

The latter theological perspective belongs to Indic traditions and religions wherein gods are intimately embedded into Nature. This vision is pantheistic both concerning the larger Hindu system of beliefs and the quasi-non-theistic Buddhist religion. Here, the material world/nature does not represent the object upon which the cultural, divine, or human subject exerts its authority, but a pervasive reality in which the subjects and the objects exist and act. In general, subjectivity may be understood as a non-transcendental energy which/who is closely linked to objectivity. The Indic view is more related to material ecocriticism with its coined term "bodymind" (Oppermann in eds. Iovino, Oppermann 29), which connects the human subject to the environmental object, the cultural body to the natural bodies, the subjective and non-transcendental mind to the immanent material bodies, both cultural and natural. In this view, food, as the environmental object of thought, is consubstantial to the human being, as the subject who sees and tastes the food.

Combining Ecocriticism, Material Ecocriticism and Ecotheology, we have come to the concept of Material Ecotheology, an incipient field of research which merges the Christian panentheistic vision with a Hindu/Buddhist pantheistic vision. It reveals the paradoxical situation in which transcendence is embedded within immanence (Osoianu 232). This postmodern perspective is in an indirect opposition to the classical, traditional and systematic-dogmatic view of God creating and governing the created world, as we can see in the Old Testament:

> Then God said, "Let Us make man in Our image, according to Our likeness; and let them rule over the fish of the sea and over the birds of the sky and over the cattle and over all the earth, and over every creeping thing that creeps on the earth." (Genesis 1:26)
>
> The fear of you and the terror of you will be on every beast of the earth and on every bird of the sky; with everything that creeps on the ground, and all the fish of the sea, into your hand they are given. (Genesis 9:2)
> You make him to rule over the works of Your hands; You have put all things under his feet (Psalm 8:6)
>
> You have put all things in subjection under his feet. "For in subjecting all things to him, He left nothing that is not subject to him. But now we do not yet see all things subjected to him. (Hebrews 2:8)[14]

As we can see, from a traditional perspective, because Man was created in God's image, he/she received the quality of ruling the creation. Many traditionalists see the act of creation as a premise for the dominion of Man over Nature. In this sense, one can conceive a binary opposition between the human subject and the natural object of domination, between the epistemic human mind and the ontic environment which gravitates around Man, between the superior godlike creation and the inferior natural creation, between the consumer and the consumed, or between the inside of the human body and the outside of the objectified food. Nevertheless, these inherent oppositions may be overcome if we think that Logos/God inhabits Nature through the agency of His homogeneous energies, which are consubstantial to Him. Thus, we can conceive Man as an image of the transcendent God and Nature as an image of the immanent God. Both types of images have a divine source and, if continuing the reasoning, the opposition Creator – creation still remains, but the opposition Man-Nature is abolished.

As stated in the Bible, the human being is made in God's image as an act of volitional and powerful creation. Nevertheless, there is the word "our" which shows that divinity in not an amorphous substance, but a polyphony of divine voices. These prosoponic voices highlight the fact that the divine substance has an ecstatic/existential form (even though transcendent and apophatic), meaning that the triadic hypostases exit the divine essence and manifest into the material

---

14 Al the quotes from The Old and New Testament are taken from the online version of the Bible https://www.biblegateway.com.

world through the agency of communion or love. Thus, the incipient premise for the creational act of volition is the love between the three divine entities. This love is translated into the generic Cosmos/Nature through two important categories of images.

First of all, in a chronological order, Nature is created as an image of the homogeneous energies of God. In other words, the natural elements and nature, in general, represent the imagistic translations of the divine logoi and, from a material ecotheological point of view, the transcendent Logos inhabits the material world by becoming immanent to it. Secondly, the human being is the prosoponic image of Logos and this perspective also implies that the human energies are images of the logoi. This "logical" chain of creational images compels us to believe that the relationship between the human subject and the natural object should reflect the relationship of love between the triadic beings. Logos is the source of the creational logoi and the chain of creational images, meaning God – Man – Nature, shows us that there is a material continuum of the human being and the environment. There are no inner hierarchies of power of the human mind over Nature, or of the human consumer over the consumed food. Food ought to be understood as a by-product of the environment, meaning that the human being has transformed the natural space into food through multiple processes of production and reproduction. The act of consuming food ought not to be viewed as an opposite of the ontological chain of creating Nature and Man in God's image. This way of embracing God's creation can be inferred from the following lines from Isaiah:

> The wolf shall live with the lamb, the leopard shall lie down with the kid, the calf and the lion and the fatling together, and a little child shall lead them.
>
> The cow and the bear shall graze, their young shall lie down together; and the lion shall eat straw like the ox.
>
> The nursing child shall play over the hole of the asp, and the weaned child shall put its hand on the adder's den.
>
> They will not hurt or destroy on all my holy mountain; for the earth will be full of the knowledge of the Lord as the waters cover the sea. (Isaiah 11: 6–9)

These Biblical lines, which almost sound like a poem, display a bucolic landscape in which all the opposites are no longer divorced, but dwelling in a paradisiac state of happiness before sin. The idyllic life in Paradise coincides with the eschatological and soteriological state of peace both in terms of the human and of the animal kingdoms. Here, Man remains the steward of Nature, but as a priest who celebrates the sanctity of all life forms, not as a cruel producer of resources and consumer of environmental products. The cosmic order is reestablished as we can see that the act of consumption or eating food does not represent an affirmation of power or hierarchy. Instead, the epistemic subject preserves the integrity of the ontic object of virtual production and consumption. The onto-epistemological imagistic relationship between God, Man and Nature shows us that the ecologies of the divine and the human subject are compatible with the environmental object or with other subjectivities found in nature (McLaughlin 113). The importance of this intrinsic compatibility also resides in acknowledging that animals to be consumed are autonomous subjectivities like the human subject and creational images:

> To think of the imago dei as a lush and divine wilderness finally is to encounter our incarnate bodies as vibrant and moving thickets of divinity, vibrant ecologies of flesh immanent yet wildly eluding creaturely feel or comprehension. The wilderness of animal bodies is a "vibrant matter," […] The creaturely bones, organs, neurons, and other companion creatures with our fragile bodies explode with an agency beyond our control. Parasites and microbial organisms mysteriously occupy our body with the fierce love of survival. The Spirit's "living profusion" of animal bodies is ever-becoming apophatic, unsaying itself in unexpected ways, in unexpected porosities (McLaughlin 89).

From this "zootheological" (88) perspective, the animal as an individual organism represents an imagistic subjectivity which displays all general features of a human being, maybe without a reflexive rationality. As previously inferred, the subjective human mind is not transcendent to the cybernetic system of Nature, but consubstantial to it. In other words, the mind, as a supposed center for subjectivity, is not superior to the object of thought, meaning nature in general and animals in particular. There is no such essential mind superior to the material body, but a holistic and integrated cybernetic bodymind who/which exists within an onto-epistemological continuum with the material world.

This pure immanence does not remove the opposition of identity-alterity in terms of perception. There is a paradox when thinking that we transcend the exterior materiality and the interior body by conceptualizing them and, at the same time, we are ontologically immanent to everything, because materiality is pervasive. The perception of transcendence needs to be corroborated to an apophatic movement of perception, when realizing that we are not identical to ourselves; besides the inner self, all the perceived materiality is alien (nature, minerals, plants, animals, our bodies, bacteria and microorganisms within us etc.). On the other hand, the perception of immanence needs to be corroborated to a cataphatic movement of perception when thinking that our bodies are identical to all material bodies at a quantic level. This weird perspective, also identified in Object-oriented ontology, leads us to reconsider our perception towards a hypothetical preferential image of God and to accept the post-epistemic/ontic interplay of identity and alterity as what we are and what we consume.

Material ecotheological considerations – An eucharistic perspective

For a better understanding of the relationship between the consumer and the consumed, we may think of the sacrament of the Eucharist, in which the bread and wine undergo an ontological (for others, symbolical) transmutation. The bread and wine become the flesh and blood of both Man and God, because Christ, through hypostatic union, shares divine and human features at the same time. In this sense, there is a metabolic circulation of logoi and images between the divine source and the creation, both human beings and Nature. The consumer (the human subject), through the sacramental act of eating, becomes one with the consumed (the natural object), meaning that the creational distance or space between the epistemic mind and the ontic environmental body is abolished. The human being becomes a bodymind, perfectly integrated within the cybernetic system of the divine and Nature:

> From these sophianic and eucharistic perspectives, food is not severance, nor does it bring about ultimate destruction and final death. Instead, what Sophia and the Eucharist convey is food as a material – as much as it is also a divine – sign of relationality, interdependence, and sharing of life eternal. The eucharistic banquet tells a story of the enactment of the Body of Christ that shapes and nurtures communal life [...] a theopolitics of Christ's Body in the Eucharist is rooted not exclusively in power, but, in a more primary sense, in divine caritas, which is expressed with a radical gesture of kenosis, reciprocity, and concrete communal practices (Méndez Montoya 114–115).

Apart from the patriarchal and Ouranic prosoponic configuration of God, the sophianic one involves two possible meanings. First, it relates to Logos and his wisdom, as seen in the doctrine of the logoi of creation. Second, it relates to a Gaian non-patriarchal vision of God inhabiting the material nature. This feminine perspective brings us closer to a more friendly attitude towards the relationship between the human being and the environment in which one lives. Man is more than a steward of Nature, as the image of the authoritative and transcendent God, acting more as an immanent priest who celebrates the miracle and sanctity of life. As a subject, the human being ought not to exert power over the environment, but to share his/her own materiality with other subjectivities, including the natural other.

Alterity, as the perceived outside of the human subject, needs not to be merely seen as a virtual product to be consumed, but an extension of one's body. Through kenosis, the image of God, both Christ and every human being, becomes empty of his/her inflated subjectivity in order to gain substance and a renewed subjectivity within the agapic and material community of Man and Nature. A new theopolitics is inaugurated when the body of Christ, both human and divine, becomes the body of the ecumenical world, including humans and all the environmental subjectivities. Concerning the bodily interplay of identity and alterity, culture and nature, divine and human, human and non-human, eating food represents a sacramental act of communion:

> When eating is Eucharistic the salvific reality of Christ is extended and made incarnate in the world. When Jesus broke bread and shared the cup as the giving of his own body and blood, and then asked his followers to "Do this in remembrance of me," he instituted a new way of eating in which followers are invited to give their lives to each other, to turn themselves into food for others, and in so doing nurture and strengthen the memberships of life. Coming to the Eucharistic table, eaters are encouraged to learn that they do not need to eat only to their own benefit and glory. They discover what is practically required to share in God's reconciliation with and within the world. (Wirzba 153–154)

The act of food eating has obvious soteriological implications; the material bread and wine represent the consubstantial image (even though the human perception and logic are incompatible with this paradox) of the material Christ's flesh and blood. Furthermore, the body of Christ represents the image of Logos which is identical to God. In this sense, Christ is both the image of God and the divine source of this image, meaning that he is consubstantial both to the divine and to the world – human and environmental. The essential and/or energetic abolition of boundaries between the creator and creation, divine and human, human and non-human, unbodied nature and material bodies makes the human being reconsider the nature of matter in general. The materiality of the world is ecumenical and ecological in terms of epistemologically breaking the limits of perception.

Considering that the act of consumption is no longer perceived as a way of exerting power over the non-human object, one may conceptualize eating as a pervasive movement of the self towards the other. Giving one's life for the other, the human subject undergoes a kenotic sublimation in terms of becoming food to be eaten by alterity. In this sense, from a material ecotheological point of view, there is a balance of exerting agentive power between the epistemic subject and the ontic object. The human being consumes the materiality of nature and, at the same time, the environmental others (human and non-human) consume him or her. Eating God's body implies eating and letting be eaten by other subjectivities. For a better understanding of the Eucharist, the text referring to the Lord's Supper (Matthew 26:26–28; Mark 14:22–24; Luke 22:17–20; 1 Corinthians

11:23–25) ought to be corroborated to the one referring to the command of not eating the forbidden fruit:

> And the Lord God commanded the man, saying, Of every tree of the garden thou mayest freely eat: But of the tree of the knowledge of good and evil, thou shalt not eat of it: for in the day that thou eatest thereof thou shalt surely die (Genesis 2:16–17)
>
> While they were eating, Jesus took a loaf of bread, and after blessing it he broke it, gave it to the disciples, and said, "Take, eat; this is my body."
>
> Then he took a cup, and after giving thanks he gave it to them, saying, "Drink from it, all of you; for this is my blood of the[a] covenant, which is poured out for many for the forgiveness of sins." (Matthew 26: 26-28)

The imperative command discriminates between the moral categories of good and evil, but this binary opposition does not belong itself to the material world, because nature is good per se. Yet, the opposition relates to the lack of maturity of the first mythical human beings, who did not have a developed moral compass. The allegorical command shows us a man and a woman who lacked the appropriate wisdom to exert free will, to possess sexual knowledge and maturity. Of course, the interpretation might be different in terms of the Promethean interdiction of the divine knowledge and the patriarchal source of this forbidden behavior. The tree of knowledge of good and evil ought to be interpreted as different to the tree of life. The latter had the role to reinforce the consequences of eating or not eating from the former one. In other words, if the first human beings had preserved intact the image of God in them by not falling (original sin), they would have gained eternal life. Instead, by consuming the fruit from the tree of knowledge and not consuming the fruit from the tree of life, their death became temporary, not identified with nothingness.

The tree of life and the possibility of resurrection are linked to the symbol of the Holy Cross, through the agency of which the material body of Christ resurrected the whole body of Nature. This process of resurrection is continuously re-enacted through eucharistic consumption, wherein the divine-human body is eaten and then it lets the possibility for the human eaters to be shared or eaten by other, within an ecumenical and sacramental liturgy. The pre-Protestant theologies consider that the presence of Christ is real, not symbolical and this presence equals to a real sharing of His body. The divine energies are bodily shared and communicated to other material bodies.

Conclusion

God's image and the human being image shares both environmental and divine features through the act of consumption. The natural ontological object does no longer gravitate around the human epistemological subject due to the fact that these apparently opposites are reconciled. The previous divorce is canceled through the volitional act of eating. Thus, the relationship of the "knowing" consumer and the "being" consumed is reinforced by the ethical "doing." The material ecotheological perspective underlines the fact that the relationship between the human subject (as consumer) and the environmental object (as the consumed) is onto-epistemological-ethical. Thus, the act of eating might be understood both as sacramental and ethical. There is a consubstantial continuum of material agentive texts, whether we are speaking about the human eater, the environmental eaten, the transcendent and immanent (at the same time) divine logoi. This pervasive material discourse reveals an ethical deletion of pre-determined epistemic and ontic boundaries between the human subject, who eats by reproducing the natural space, and the natural object, which is eaten and embedded into the human bodymind.

References

Heynen, Nik. Kaika, Maria. Swyngedouw, Erik. *In the Nature of Cities: Urban Political Ecology and the Politics of Urban Metabolism (Questioning Cities)*. London: Routledge, 2006. Print.

Iovino, Serenella. Oppermann, Serpil. *Material Ecocriticism*. Bloomington: Indiana University Press, 2014. Print.

Iovino, Serenella. Oppermann, Serpil. "Material Ecocriticism: Materiality, Agency, and Models of Narrativity". *Ecozon@: European Journal of Culture, Literature and Environment*, 3.2 (2012): 75–91. Web.

McLaughlin, Ryan Patrick. *Christian Theology and the Status of Animals: The Dominant Traditions and Its Alternatives*. London: Palgrave Macmillan, 2014. Print.

Méndez Montoya, Angel F. *Theology of Food. Eating and the Eucharist*. Malden: Wiley-Blackwell, 2009. Print.

Moore, Stephen D. eds. *Divinanimality. Animal Theory, Creaturely Theology*. New York: Fordham University Press, 2014. Print.

Oppermann, Serpil. "Material Ecocriticism and the Creativity of Storied Matter". *Journal of Literary Studies* 26.2 (Ecocriticism Special Issue) (December 2013): 55–69. Web.

Osoianu, Dragoș. *Urban Ecocriticism and T.S. Eliot's The Waste Land*. Craiova: Universitaria, 2018. Print.

Yannaras, Christos. *Person and Eros* (trans. by Norman Russel). Brookline: Holy Cross Orthodox Press, 2007. Print.

Wirzba, Norman. *Food & Faith. A Theology of Eating*. Cambridge: Cambridge University Press, 2011. Print.

The Bible. https://www.biblegateway.com, n.d. Web.

Part II: Eco-Consciousness and Eco-Activism

Adina Ciugureanu

Remapping *The Waste Land* and Climate Change

> And I will lay it waste; it shall not be pruned, nor digged; but there shall come up briars and thorns: I will also command the clouds that they rain no rain upon it.
> *Isaiah, 5:6*

Abstract: The present chapter aims at re-visiting T.S. Eliot's major poem, *The Waste Land* (1922) through the lens of eco- and geo-criticism, two of the more recent critical approaches. I argue that the combination of the two perspectives (reading the poem as a prophecy of the possible apocalyptic death of the earth by lack of water and as a geographical mapping of London in the aftermath of WWI) reveals both Eliot's modernist reading of urban space and his knowledge of the climate change views of the time.

Introduction

Eliot's allusion to Isaiah's prophecy in *The Waste Land* has already been documented. It has also propelled the poet to the position of a prophet of the modern age. Though it has been one hundred years since the publication of the poem, its significance in the recent debates on climate change and alleged imminent dryness of the inhabited areas of the globe makes it worth revisiting from both eco- and geocritical perspectives.

No one of those familiar with Eliot's poem can forget the poet's description of "land" as "dead," "stony," "a handful of dust," "desert," "only rock," "cracked," "arid," which I have interpreted as "objective-correlatives" for "infertility," sterility, and death (Ciugureanu 124–125). But Eliot's unforgettable depiction of the land also fits into the grand image of the apocalypse, one of the major concerns of ecocritical studies (Garrard 2004). Apocalyptic literature or apocalypticism (as it is also called) has been embedded in Euroasian culture for about four millennia, starting with the Old Testament and continuing with the apocalyptic revelations contained in the New Testament. Apocalypticism has a discourse of its own, meant to unveil (Gr. *Apo-calyptein*) the moment when the world as we know it comes to an end, caused by a deep spiritual and environmental crisis. Described as the end of time and the end of history, apocalypticism responds to a crisis that affects both nature (the environment) and human nature (mankind) because the two cannot be separated. Therefore, apocalypticism is looked at as an important ecocritical trope, which is worth exploring (Garrard 86).

Against the background of arid lands and plains, Eliot's introduction of, and return to, what he calls "Unreal City" (Parts II, III and V) show his obsession with urban areas, which he envisages as images of modernity, but also of disillusion, of fragmentation, and destruction. Whatever the city represents to the poet, it clearly emphasizes the existence of space and place which could be mapped, but also configured in relation to the people living in them and to the environment. A cartographic representation of Eliot's "Unreal City" reveals real places that compose the city, divided between the mundane (the profane) and the sacred, encompassing what St Augustine, one of Eliot's inspirations for the poem, calls the City of God as opposed to the City of Man.

By mapping Eliot's city, I intend to disclose and explore the significance of space and place in Eliot's cultural, but also geographical reading of the world. In his way, Eliot creates both a textual geography and a literary cartography of the western civilization in the aftermath of WWI. On the other hand, space is known to be configured by environmental markers, which could be private (a room), public (a pub, a church) general (borders, lands, mountains, rivers) or even climate zones (deserts, oases). The exploration of these markers from the perspective of the increasing volatility they are threatened by, of crises in the social and natural worlds they represent, reflects an ecocritical concern. As Tally and Battista point out, bringing together literary studies and ecological concerns, laying emphasis on space, place and mapping in the Anthropocene age we have been living in for at least three centuries, allows for a deeper analysis of "the fractured, disjunctive spatial anxieties of modernity or postmodernity," with a view to "simultaneously imagining a more sustainable modality of environmental inhabitation" (8). Therefore, a joint ecocritical and geocritical approach offers lines of inquiry that examine the intersections between mapping literary spaces and exploring ecological consciousness. This is the line of reading which will be followed by the revisitation of *The Waste Land* through re-mapping its spaces and reconsidering the apocalyptic scenes which the poem reveals. That Eliot has been influenced by the biblical discourse of destruction has already been discussed by critics (Coote, Kermode, Kenner, among others). That Eliot is himself a prophet for an incoming apocalypse which is not only spiritual, but also as physical as possible and that at this very moment we may be approaching a climatic apocalypse are questions this essay intends to answer.

Revisiting the "Unreal City"

In *The Book of the Prophet*, Isaiah reveals how the Jewish people will be castigated (by God) for their backsliding and lack of faith in the Creator with a rhetoric of apocalyptic destruction: "And I will lay it waste; it shall not be pruned, nor digged; but there shall come up briars and thorns: I will also command the clouds that they rain no rain upon it." (Isaiah 5:6)[15] Described, metaphorically, as God's vineyard, the Jews are menaced to be laid waste, infertile, and returned to the parched desert. Unless they change, they will be denied the promised land. Inspired by Isaiah of the Old Testament, Eliot also suggests an imminent apocalypse for the unfaithful and unrighteous which will scorch the land they live on.

> What are the roots that clutch, what branches grow
> Out of this stony rubbish? – an allusion to vines (ll 19–20)

and

> Here is no water but only rock
> Rock and no water and the sandy road
> The road winding above among the mountains
> Which are mountains of rock without water
> ………………………………………………..
> Amongst the rock one cannot stop or think
> Sweat is dry and feet are in the sand –

While in lines 19–20, the allusion to dry vines reveals the deadly desert which God's vineyard could turn into, lines 331–337 remind the reader about the Exodus and the Jews' retreat into the mountains in various times of their history as a punishment for their infidelity.

In the *Book of Revelation*, another Biblical text that had a strong influence on Eliot's poem, the city of Jerusalem (or the New Jerusalem) appears as "the holy city [...] coming down from God out of heaven" (Rev. 21:2), while in *Hebrews* it is described as "the city of the living God, the heavenly Jerusalem" (Heb. 12:22). St Augustine, however, envisions the city as two contrasting urban spaces: one is the "City of God" ("the holy city") while the other one is the "City of Man," of

---

15 In the Bible on line version, the translation of Isaiah's words is even more revealing of the apocalyptic description of the world's end than in King James's version of the Bible which Eliot definitely had read and which I used in this essay. "I will make it a wasteland, neither pruned nor cultivated, and briars and thorns will grow there. I will command the clouds not to rain on it." (Isaiah 5:6; https://www.biblegateway.com/passage/?search=Isaiah%205-6&version=NIV)

Babylonian structure (O'Daly 53–54). To Augustine, the City of God is spiritual, it is a goal to achieve, whereas the City of Man is physical, terrestrial, therefore, sinful and egoistical. In geocritical terms, the City of God could be described as imaginary or virtual, while the City of Man can be looked at as the built-in environment in which people live. The connection between the two kinds of city, which could co-exist in the same urban space, is created by *civitas dei*, the God loving community of people, who aspire to salvation, but who are constantly confronted by the pagan community of the people of Babylon, doomed to eternal damnation. The city, its space and places, becomes, therefore, a site of tensions and negotiations, of salvation and damnation, of the sacred and the profane.

Eliot's "Unreal City" could be definitely read as a site of tension between salvation and damnation, a City of Man, trying hard to aspire to the heavenly City of God. "Unreal City" appears for the first time in "The Burial of the Dead." It is a cold city ("winter") covered with "brown fog," very early in the morning ("dawn"). The geographical point of reference in this "unreal city" is "London Bridge," over which a crowd is flowing

> .... up the hill and down King William Street,
> To where Saint Mary Woolnoth kept the hours
> With a dead sound on the final stroke of nine. (ll. 66–68)

Leaving aside the cultural allusion to Dante and his description of the flow of dead people walking slowly, eyes downcast, to the Inferno, Eliot describes an ordinary morning when the people working in the City of London cross the bridge. He was one of them because he was working as a bank clerk at the time and, every morning, he crossed London Bridge to King William Street, passed by Saint Mary Woolnoth on his way to the financial district and Lloyds Bank.[16] The stroke of nine, marking the start of work, stresses the connection between the church and the City of Man in which a sacred place, the church, loses its meaning of the house of God and becomes a ritualistic marker for mundane living.

"Unreal City" reappears in "The Fire Sermon" part and displays more scenes of a modern, yet boring, sterile urban life. The City is covered with the same brown fog, it is still winter, but the time is noon, more specifically, lunch time. The first two points of reference are close to the financial district. Mr. Eugenides, "the Smyrna merchant," invites the speaker to lunch at the Cannon Street Hotel

---

16 Eliot took the position of accountant at Lloyds Bank in 1917 and left it in 1925.

and then to spend the weekend at the Metropole. Later, "at the violet hour," when work ends, a typist, "home at teatime," waits for her lover, "a small house agent's clerk," with whom she shares the meal and the bed. Both are bored, tired and indifferent to each other. The music on the gramophone she starts listening to after her boring lover left flows "along the Strand, up Queen Victoria Street," while "in lower Thames Street fishmen "clatter and chatter" in a pub near "the walls / Of Magnus Martyr" which "hold[s] / Inexplicable splendor of Ionian white and gold." (ll 262-265).

The mention of St. Magnus Martyr, built in the second half of the eleventh century and restored by Christopher Wren in the seventeenth century, is probably due to the fact that in 1920 the church was on the list of the nineteen London churches to be demolished. Eliot joined the group who rose against this decision, writing that St. Magnus Martyr bestowed beauty on "the business quarter of London [with] its hideous banks and commercial houses" and that "[t]he loss of these towers, to meet the eye down a grimy lane, and of these empty naves, to receive the solitary visitor at noon from the dust and tumult of Lombard Street, will be irreparable and unforgotten" (in Spurr 36). The church was eventually saved, but Eliot who was at the time on his way to converting to Anglicanism, most probably felt the need to mention its sacred beauty and religious meaning in *The Waste Land*.

Like Saint Mary Woolnoth in "The Burial of the Dead," St Magnus Martyr is reminiscent of the City of God, which is threatened to lose its meaning (the spire clock announces the start of work, not of prayer, the towers of the house of God are about to be demolished) and perish in the City of Man suffocated by work, boredom, superficiality, lack of faith or of any other kind of spirituality. The juxtaposition of, and tension between, the places in which common people (secretaries, agent's clerks, merchants, fishmen) work and socialize and the places of devotion in which people should, at least, keep their connection with the Christian God map not only a modern city, but also a post WWI apocalyptic space where air impurity matches arid spirituality, just as nature falls prey to human activity in industrial cities.

> The barges wash
> Drifting logs
>   Down Greenwich reach
>   Past the Isle of Dogs (ll 272-275)

Geocritical studies (Lehan, Short, Spates & Macionis) refer to the paradigm shift from the spiritual to the commercial city, from the artisan-minded to the industrial city at the dawn of capitalism when the feudal system is gradually replaced with the capital-based society and when the modern city starts growing as a result of urbanization. Eliot's city is no longer the modern city in its beginnings, but the city that has known modernity and is now at its nadir and falling apart. It cannot contain itself any longer.

The City of Man

The population of the modern city consists of the crowd, in which individuals are perceived to look almost identical, and the *flâneur*, the stroller, who detaches himself from the crowd and acquires particular features. In Eliot's Unreal City,

> A crowd flowed over London Bridge, so many,
> I had not thought death had undone so many.
> Sighs, short and infrequent, were exhaled,
> And each man fixed his eyes before his feet. (ll 62–65)

The crowd, whose composition suggests commoners and working people, is referred to again in "The Fire Sermon" when it is described as a "human engine," "the eyes and back / Turn[ing] upward from the desk, when the human engine waits / Like a taxi throbbing, waiting" (ll 215–217).

Interestingly, "human engine" is one of the terms used in apocalyptic narratives, in the virtual and augmented reality of video games, and is included in the more recent concept of transhumanism (cyberbodies). In "What the Thunder Said," the crowd has turned into "hooded hordes swarming / Over endless plains, stumbling in cracked earth / Ringed by the flat horizon only" (ll. 368–370). The "hooded hordes" prefigure the "hollow men" of the poem with the same title, published three years later, who are depicted as eyeless (their "eyes are not here").

> There are no eyes here
> In this valley of dying stars
> In this hollow valley
> This broken jaw of our lost kingdoms.
>   (*The Hollow Men*)

While the indistinguishable people forming the crowd look more and more lifeless than alive, the individual characters (Lil and her friend, the neurotic woman, the typist, the clerk's agent) share with the crowd their deserted souls.

Against this background, the "I" of the poem, the *flâneur*, is a chameleonic self, changing masks and moving fast among places both horizontally (geographically, in space) and vertically (historically, in time). The Unreal City map expands from the financial district, London Bridge and the Lower Thames docks to the Greater London area: Highbury and Moorgate (situated north of the financial district), Richmond and Kew, south-west of London, and from Greater London to the Channel (Margate Sands). The "heap of broken images," the scenes and fragments that compose the poem send the reader to diverse cultures and different historical periods, all illustrated by, and connected to, an arid, infertile land, on the verge of total collapse and irremediable destruction.

The journeys across the waste land are both horizontal (from London to Vienna, Venice, Jerusalem) and vertical: from London to ancient Rome (alluded by Carthage and the Punic war), to ancient Greece (Athens, Alexandria), and ancient Jerusalem. What these cities had in common at some point in their history was the "falling towers" (l. 373). Like ancient Babylon, they ended in complete ruin. Three of these cities (Carthage, Vienna, and London) were capitals of famous empires that either faced destruction in the past or were facing fragmentation and decline in Eliot's time. The Phoenician world was destroyed by the Romans, the Roman colonial world ended in ruin, being followed by the Holy-Roman Empire which allowed for the Austro-Hungarian Empire to emerge in central Europe as its heir, and, finally, the British Empire. While the Romans and their colonies were a thing of the past, the Austro-Hungarian Empire declined after WWI, losing part of its territories, while the British Empire which was, apparently, still enjoying its zenith, was soon to fall apart.

Eliot's prophetic depiction of an apocalyptic London prefigures the irreversible fragmentation of the British Empire in the twentieth century. Any cartographer would see the mapping of these three empires as a palimpsest of ancient and modern places, religions, historical figures and events and, consequently, as a constant remapping of the world. Therefore, Eliot's poem is not only about London and arid lands, but also about mapping and remapping geographical and cultural spaces. According to Eleanor Cook, Eliot's poem involves three maps: one of a city with London as the center, alluding to the British Empire as well, one of an empire with Rome as the center, and one of a world with Jerusalem as the center which overlaps and illuminates the others (Cook 341–343). In this way, Jerusalem that had known its fallen towers becomes the hope and the light for the holy and heavenly city, for the City of God, in St Augustine's words.

The city of god

Alongside the images of real and mythic cities mentioned above, the poem moves outside of western culture, to India and the myth of the thunder god, Prajapati. The Hindu myth juxtaposes the Egyptian fertility myth and the Grail legend. They all have one point in common: barrenness, scorchedness, infertility caused by the absence of one of the four primordial elements of nature: water.

By comparing the words which Eliot uses to suggest the absence of water (16 in number) with the words used to suggest arid land (11), destructive fire (6) and unbreathable air (10), one can notice that the stress falls on water, one of the primary elements in the absence of which life does not exist. Eliot's cosmology lacks the equilibrium it should have and, as a consequence, the imbalance created leads to the sickness and death of both the environment and human nature. Seen from this perspective, the crowd people, the individual characters and even the chameleonic self who narrates the "heap of broken images" make sense. From the "dead land" (line 2) to the "arid plane" (line 424), the reader crosses the dryness of deserts and mountains of rock with the exception of a small oasis: the "hyacinth garden," a symbol for the resurrected god of the fertility rites, too weak to perform any renewal of the land. Water has a double meaning in this context: it is both an element of fear ("Death by Water") and the basic fertilizer of the land. The absence of water brings about both the barrenness of the land and sexual and spiritual sterility. Since water can no longer feed the earth, the air becomes corrupt (brown fog, violet air) and the fire is allowed to destroy ("Burning burning burning burning/O Lord Thou pluckest me out/O Lord Thou pluckest/Burning," ll 308–311).

The provider of water, in Eliot's view, is God for the western world and Prajapati for eastern culture. Both produce and have produced thunder throughout mythical history.

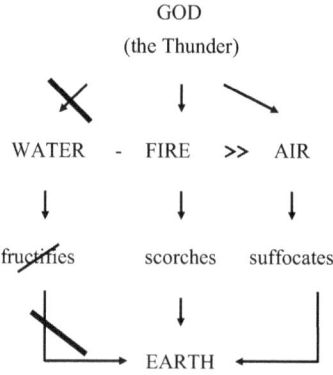

Fig. 1 (Ciugureanu 124)

The broken link between water and earth, the effect of God's refusal to reset the equilibrium, or his inability to re-create it, obviously cause a disruption between the other elements as well. The result is "a heap of broken images," apparently unconnected to each other in the texture of the poem. However, the symbolic representations of the four elements are the "objective correlative" that expresses the poet's feelings of desolation, futility and artificiality regarding both nature and man. They make up the patchwork of the Eliotesque urban landscape raised to universal value.

Contrary to the dire description of the waste land throughout the text, the ending of the poem could be read as an optimistic note when both the Fisher King and the Thunder God make their appearance. The Fisher King, identified with "the man with three staves" (l. 51) in "The Burial of the Dead," reappears in "What the Thunder Said" ("I sat upon the shore/Fishing," ll. 423–424) but merges into the figure of Christ, the Fisher of men. Prajapati, the intelligent heart of the universe, is introduced in the last part of the poem through his voice by means of which he communicates to the world. What he says is: "Da." The Sanskrit word, "Da," is differently understood by the hearers, according to their flaws. Therefore, the gods, who are naturally unruly, take "Da" for "Damyata" (control or surrender yourself); men, who are generally avaricious and selfish, think that the Thunder said "Datta" (give), and demons, who are usually hostile to people, understand "Dayadhvam" (be compassionate). It is these three magic words that, unless turned into action, will not persuade the Thunder to make rain fall. The poem ends with another Sanskrit word, the mantra: *shantih* (peace beyond understanding).

Although the repetition of "shantih" is meant to appease the tensions created throughout the text, it raises a few questions. Does it suggest the return of fertility and life or the breaking loose of the apocalyptic end of the world? Will the three groups of hearers follow the Thunder's command or will they ignore it? Is this "peace beyond understanding" that Eliot invokes a fresh beginning for the world and the recovery of land or does it suggest the calmness that precedes a destructive storm that may wash away all civilizations?

Waste land and climate change

In any culture, there has always been an interest in the health of land, in its fertility and usefulness. The famous American ecologist, scientist, and naturalist Aldo Leopold, for example, who wrote about ways of controlling the health of land almost a century ago, describes the causes that may make soil lose fertility: "It is now generally understood that when soil loses fertility, or washes away faster than it forms, and when water systems exhibit abnormal floods and shortages, the land is sick" (in Williams 449).

The sickness of land, whose cause is water flooding or drought, by ineffective technology, by adverse human activity, by natural causes, by climate change, having desertification as a main effect. In ecological studies, there has always been an interest in the life of deserts (their birth, change, growth or decline), especially in the last two centuries, when the issues of climatic variations appeared. Thus, it has been demonstrated that human activities largely contribute to floods and droughts in the desert world and its margins, to desert expansion and desert formation in areas which had been fertile in the recent past (Williams 450). In all these cases the element which is essential in keeping the soil healthy or in creating deserts is water, which may destroy land both in the form of floods (when in excess) and in the form of droughts (lack of water). The concept of "desertification," used by ecologists to describe the process of desert formation and its subsequent changes has been recently defined by ecologists. In his 2014 study, *Climate Change in Deserts (Past, Present and Future)* Martin Williams analyzes all the points of view and describes desertification as a process caused both by adverse human activities and by climatic variations, the latter being considered by most ecologists to be the primary cause of desert expansion (473–474). Prolonged droughts and sustained lack of rain are considered by scientists to have contributed to widespread land degradation (474).

On the other hand, desertification may also be connected to global warming, which became a research topic as early as the start of the nineteenth century when the effects of industrialization began to be known. The nineteenth century scientists measured the release of carbon dioxide (CO_2) into the atmosphere and came to the conclusion that the increase in hydrocarbons and CO_2 leads to a rise in temperature on the surface of the earth.[17] In addition, too many uncontrolled gases released into the atmosphere lead to the production of what is called the "greenhouse effect," a term used for the first time in 1901[18], which soon became a household word. The dry seasons at the start of the twentieth century almost convinced the scientists that a period of climate warming was following and comments on how coal burning at a large scale and the continuous emission of

---

17 The discovery of the processes that lead the greenhouse effect in the atmosphere was made by the French Joseph Fourier (1824) but the analysis of the effect of various gas emissions into the atmosphere was first quantified by the Swede Svante Arrhenius in 1896. He was also the first to predict global warming.

18 The term "greenhouse effect" was used for the first time by the Swede Nils Gustaf Ekholm to describe the effect of carbon dioxide to produce a new ice age, unless it is controlled.

carbon dioxide and other harmful gases into the air were affecting the climate. Articles in various scientific journals were published at the time.

Eliot's knowledge of chemistry, at least of gas combinations, is well known. In *Tradition and the Individual Talent* (1918) his reference to the formation of sulphurous acid when he compares the gas combination with the creation of the work of art by the poet's mind is famous. There is reason to believe that Eliot was not unaware of the climatic variation narratives of his time.[19] The *Popular Mechanics* magazine, for instance, published in the United States was very popular on the continent as well. It contained articles about the possibility of global warming or cooling as a result of the changes in atmosphere. Both predictions for the twentieth century had as an effect the infertility of land and/or desert formation. One of these articles by Francis Molena, published in 1911, does an interesting and informed analysis of climate changes and its cycles throughout time and, by referring to the 1911 atypical weather, it claims that

> There are indications that the maximum of the warm era in which we now live has not yet been reached and that, therefore, the climate will probably become slowly warmer for some thousands of years, this progression being broken by many small oscillations. (Molena 340)

The cause of the early twentieth-century global warming prediction was the rise in carbon dioxide percentage contained in the air, through the combustion of coal or vegetable material, seen as an effect of industrialization and a producer of the greenhouse effect. Molena's description of the earth's atmosphere and air circulation upon which the climate is dependent and its connection to the sun that provides the heat, his mention of stormy weather in a few American states and scorched land in others frame a journalistic discourse which Eliot could have brilliantly turned into a poetic one.

Could it be possible then that Eliot's description of scorched land, drought and fire be connected to the greenhouse effect discussions of the time? Could it be possible then that *The Waste Land* may be read as a foreboding of an imminent apocalypse based not only on human destruction, but also on climate change? Interestingly enough, both the author of the article and the poet look at the imbalances of nature globally, from both geographical and historical perspectives, both in space and in time. Though they are optimistic about the future,

---

19 See Francis Molena's article, "Remarkable Weather of 1911 (The Effect ai the Combustion of Coal on the Climate – What Scientists Predict for the Future)" in *Popular Mechanics*, March 1912, pp. 339–342.

Molena considers that real and palpable changes in climate will take generations, while Eliot foresees salvation in the return to faith.

Conclusion

Eliot's capital poem has generally been looked at aa a revealing example of high modernism in poetry when the subjective "I" becomes an objective artistic experience, when personal emotion is replaced by "a set of objects, a situation, a chain of events," which will form "the objective correlative" of a particular sensory experience. (in ed Kermode 48)

Looked at from the ecocritical or eco-conscious perspective, Eliot's *Waste Land* could reveal more than it meets the eye, a deeper layer which may support the poet's depiction of the apocalyptic images not only as the disastrous effect of the war, but also as the imminent effect of gas emission caused by industrialization in the "unreal cities" of the world. The remedy for the salvation of the earth and of civilization seems to be only one: return to faith and to the reconstruction of the City of God. The key to cure the earth seems to lie in the hands of *civitas dei*.

References

Ciugureanu, Adina. *High Modernist poetic Discourse*. Constanța: Ex Ponto, 1997. Print

Cook, Eleanor. "T. S Eliot and the Carthaginian Peace." *ELH* 46.2 (Summer 1979): 341–355. Web.

Eliot, T.S. *Collected Poems (1909–1962)*. London and Boston: Faber and Faber, 1963. Print.

Kenner, Hugh. "The Urban Apocalypse" in ed. A. Walton Litz. *Eliot in His Time: Essays on the Occasion of the Fiftieth Anniversary of The Waste Land*. London: Oxford UP, 1973. (23–50) Print.

Kermode, Frank (ed.). *Selected Prose of T.S. Eliot*. London and Boston: Faber and Faber, 1975. Print

Mayer, John. *T.S. Eliot's Silent Voices*. New York and Oxford: Oxford University Press, 1989. Print.

Lehan, Richard Daniel. *The City in Literature – An Intellectual and Cultural History*. Berkeley, CA and London: University of California Press, 1998. Print.

Molena, Francis. "Remarkable Weather of 1911 (The Effect ai the Combustion of Coal on the Climate – What Scientists Predict for the Future)." *Popular*

Mechanics (March 1912): 339–342. https://archive.org/details/PopularMechanics1912/Popular_Mechanics_03_1912/page/n85/mode/2up

O'Daly, Gerard. *Augustine's City of God. A Reader's Guide.* Oxford: Oxford University Press, 1999. Print.

Pike, David L. "Modernist Space and the Transformation of Underground London," in ed. Pamela K. Gilbert. *Imagined Londons.* New York: State University of New York Press, 2002. (101–120). Print.

Rainey, Lawrence (ed.). *The Annotated Waste Land with Eliot's Contemporary Prose. With Annotation and Introduction.* New Haven and London: Yale University Press, 2005. Print.

Short, John Rennie. *Urban Theory (A Critical Assessment).* New York, NY: Palgrave Macmillan, 2006. Print.

Spates, James L. and Macionis, John J. *The Sociology of Cities.* (2nd ed.). Belmont, CA: Wadsworth Publishing, 1987. Print.

Spurr, Barry. *Anglo-Catholic in Religion – T.S. Eliot and Christianity.* Cambridge: Lutterworth Press, 2010. Print.

Tally, Robert, Jr. (ed.). *Spatial Literary Studies Interdisciplinary Approaches to Space, Geography, and the Imagination.* London: Routledge, 2021. Print.

Tally Robert, Jr. and Christine Battista (eds.). *Ecocriticism and Geocriticism. Overlapping Territories in Environmental and Spatial Literary Studies.* London: Palgrave Macmillan, 2016. Print.

Yang, Carol L. "*The Waste Land* and the Virtual City" in ed. Joe Moffett, *The Waste Land at 90. A Retrospective.* Amsterdam, New York: Rodopi, 2011. (187–216). Print.

The Bible. Authorized King James Version. With an Introduction and Notes by Robert Carroll and Stephen Prickett. New York: Oxford University Press, 2008 [1997]. Print.

The Bible. https://www.biblegateway.com/passage/?search=Isaiah%205-6&version=NIV. Web.

Thompson, D. *The End of Time: Faith and Fear in the Shadow of the Millennium.* London: Minerva, 1997. Print.

Williams, Martin. *Climate Change in Deserts (Past, Present and Future).* New York: Cambridge University Press. 2014. Print.

Alina Cojocaru

Transatlantic Perspectives on Urban Environments in T.S. Eliot's *The Waste Land* and William Carlos Williams's *Paterson*

Abstract: This chapter explores the nexus between geocriticism and ecocriticism in order to challenge the reductive vision of an urban-centered modernist writing which is disengaged from nature. It argues, instead, that both T.S. Eliot's *The Waste Land* and William Carlos Williams's *Paterson* do not only portray "wastelandic" depictions of urban environments but also awaken a collective environmental consciousness.

Introduction

Literary modernism may be perceived as an incongruous space for ecocriticism. The aftermaths of the Great War and the Second World War represented a catalyst for "wastelandic" depictions of urban environments fueled by technological advances. Michel de Certeau perceives the modern city as "the machinery and the hero of the modernity" (95), a central stimulus for creation which has been mirrored in a kaleidoscope of responses projecting a disunified vision. Modernity, with its climate of personal and cultural anxiety, ensued as a result of the transformation brought about by urbanization, secularization and industrial development. In fact, the first use of the word "modernity" by Baudelaire in *The Painter of Modern Life* emphasized precisely the fashionable, the artificial, the ephemeral from which artists captured the spirit of the moment (Childs 15), whereas nature acquired negative connotations. As war poets recounted their experiences in the trenches, modernists symbiotically espoused an overhaul of their vision on nature, time, consciousness, poetic language and form. In *The Great War and Modern Memory*, Paul Fussell stresses the difference between pre- and post-war British poetry, observing that in the poetry anthologies that soldiers carried "we find that half the poems are about flowers and that a third seem somehow to involve roses" (Fussell 291). By virtue that "there was no *Waste Land*, with its rats, alleys, dull canals, and dead men who have lost their bones" (Fussell 25), war poets unflinchingly recounted their experiences in the trenches, going beyond the lapsed artistic naiveté of glorifying war. This transfiguration of violence into creativity was embraced by modernists who subsequently adopted their own militant rhetoric through their metaphorical "slaughter" of established conventions. Poetry itself became a war casualty.

T.S. Eliot's *The Waste Land* and William Carlos Williams's *Paterson* may be separated by over two decades, yet they are arguably complementary in their perspectives on the urban environment and the "urban climates" concerning the new ideas, philosophies and politics that permeate the city on the two continents following the First World War and the Second World War, respectively. It is well-known that the poetry of T.S. Eliot and William Carlos Williams, although rooted in the same Imagist theory of Ezra Pound, expresses the tensions between the high-brow and low-brow modernism of the twentieth-century. If the former is attuned to the classical tradition, William Carlos Williams responds to the Anglocentric cosmopolitanism of *The Waste Land* by dwelling on the margins of traditions and grounding *Paterson* in American realities. The morphing city plagued by social and cultural damage constitutes a focal point of both poems. So does the threat of an environmental collapse. In this respect, instead of relying on the nature/culture, rural/urban dichotomies, I employ the representation that Claude Lévi-Strauss has of the city as "at one and the same time an object of nature and a subject of culture" (*Tristes Tropiques* 124), the concept of "urbanature" (Nichols) and "rhizomatic maps" (Deleuze and Guattari) in order to create a nexus between ecocriticism, geocriticism and the modernist city. In doing so, the modernist representations of urban environments in the two poems are reconsidered as repositories of an environmental consciousness.

The many facets of nature in *The Waste Land* and *Paterson*

Through the use of multiple perspectives, the plights of the modern world unfold in the two poems, revealing complex human/nature correlations. The term "nature" itself is rather elusive. Raymond Williams defines nature as "'perhaps the most complex word in the language,'" highlighting its meanings as the innate qualities of something, the inherent force driving all existence and the material world that unravels before our eyes (165). This effacement of the divide between nature and culture complements well the definition that Ursula K. Heise gives to ecocriticism, namely the "attempt to think beyond conceptual dichotomies […] – the separation of subject and object, body and environment, nature and culture" (50–67). Furthermore, it allows for the poetic representation of nature in all its potentialities. Nature is asserted as essential in all aspects of human experience since environmentally damaging practices are harbingers of a general collapse of civilization. In this respect, T.S Eliot employs the objective correlative to address the correlation between his inner wasteland the wasteland without, choosing "[a] heap of broken images" (*The Waste Land* 5) to represent an apocalyptic, barren wasteland, where the primordial elements are no longer in harmony. Humanity has suppressed its innate earthy connections and has therefore become devoid of a cosmological dwelling in the course of changing the natural environment.

From an external perspective, *The Waste Land* presents itself in consonance with the definition of ecocriticism advanced by Cheryl Glotfelty. Glotfelty insists on the idea of "integrated systems, and strong connections among constituent parts" (xx) of the ecosphere, in a similar manner that T.S Eliot advocates in his "Notes Towards the Definition of Culture" for "variety in unity: not the unity of organization, but the unity of nature" (*Selected Prose* 302). The mythic patterning informed by the readings of Jessie Weston's *From Ritual to Romance* and James Frazer's *The Golden Bough* constitutes a unifying element which offers centrality to rituals of vegetation. Analogously, John Elder, in another pioneering work of ecocriticism titled *Imagining the Earth: Poetry and the Vision of Nature*, states that that there is "redemption offered to human cycles within the order of natural cycles, an equilibrium as precise and comprehensive as an ecosystem" (qtd. in Philips 155). By asking "What are the roots that clutch, what branches grow/ Out of this stony rubbish?" (*The Waste Land* 5), the poem may be voicing the disconnection and dismay that characterize individual responses to environmental change. As unexpected as they may be, parallels between modernism and ecocriticism challenge the reductive vision of a postwar urban environment disengagement with nature. What is particularly fascinating, yet has been less researched so far, is precisely this reconsideration of modernist writers as relevant voices in the current discourse of environmental awareness.

There is a visionary ecology present in *The Waste Land* through the juxtaposition of disjointed agencies and spatial imagery that materialize in the context of the natural world. There is no chronological progression of the poem as it describes the scale of the anthropogenic interference. According to Buell, "what human society, at present, is extinguishing at so unprecedentedly dizzying a rate can be repaired only in time frames at least several times longer than the evolutionary span of the human species" (96). Consequently, the damage done to the environment exceeds the boundaries of human subjectivity and the individual's cosmic position. This awareness of the multiple agents that shape physical and metaphorical climates may account for the extinction of personality, the escape from emotion that Eliot endorses. The mind of the poet sifts the debris of the urban environment with scientific precision since his mind remains, as described in his landmark essay "Tradition and the Individual Talent," an "inert, neutral and unchanged" receptacle (*Selected Prose* 41). For Eliot, contemporary civilization seems not to be eligible to enforce any principles of organizing nature. The main by-product of modernity is waste.

What ensues is a confrontation with the desolate climate. Harrison argued that "the greenhouse effect, or desertification of habitat in general is the true 'objective correlative' of the poem" (Harrison 149). Naturally, one might challenge this idea since Harrison does not offer further explanations about his understanding of the objective correlative from an ecocritical perspective. For T.S. Eliot, the war and the broken images of decay in the real world, also in relation to the mythic method, thus evoking aspects of vegetation myths, fertility and infertility, constitute the objective correlative of his subjective, vulnerable self. In this respect, the poetry of T.S. Eliot, the patient, which envisages the urban tissue as a literary mirror of the physical and moral afflictions of the modern subject, diverges from the vision of doctor William Carlos Williams, that operates with the idea of de(con)struction as a harbinger of a cure for the ills of modernity rather than a disease, albeit both poems arguably strive to create order out of the chaos intrinsic to the urban environment.

In line with the apocalyptic images associated with climate change, there is a sense of crisis expressed in "The Burial of the Dead" by the observation that "the dead tree gives no shelter, the cricket no relief" (23). Furthermore, "the brown fog of the winter dawn" (25) draws the attention to the smog enveloping urban areas. In "A Game of Chess," the natural becomes artificial, but also lifeless in the process. Rats scurrying within the confines of the city and outside them, in the trenches, not only foreground an anti urban discourse but also the propagation of disease. "The Fire Sermon" depicts the pollution of water sources through a clutter of "empty bottles, sandwich papers, / Silk handkerchiefs, cardboard boxes, cigarette ends" (30). The filth of London is carried by the river into the sea, where the death by water of Phlebas highlights a cycle of corruption and regeneration. In "What the Thunder Said," the extinction of the land may be metonymically represented by the hermit-thrush that Eliot recognized as awaiting death when faced with a catastrophic drought (Tobias and Morrison 29). Land, water and air bear the signs of a visible pollution with human debris. The leitmotif of dirt and death articulate considerations of the city as a site of decay, excess and abandonment.

However, is there value in the waste of the urban environment? Reinterpreting waste as a cultural by-product offers a new outlook on the line "I will show you fear in a handful of dust" (*The Waste Land* 6). In addition to the pronounced sterility of the wasteland, a certain degree of revulsion against the material aspects of the world can be sensed. Heather Sullivan proposes "dirt theory" to reveal how in modernity "efforts to conceal 'dirt' in its many forms have encouraged urban residents to believe that dirty nature is something far away and disconnected from themselves and their bodies" (526). Dirt is connected to death, waste and the repressed, triggering images related to the unearthing of the horrors of war. Urban poverty and waste, instantiated by "broken fingernails of dirty hands" (*The Waste Land* 15), decaying teeth and "dirty" ears, come as an extension to the decomposing bodies from the war trenches. Nevertheless, the past cannot be buried and it is the role of the poet to dig what civilization is trying to omit or marginalize, akin to the dog digging "the corpse [...] planted last year in the garden" (*The Waste Land* 7). It could be argued that waste is neither disposable nor replicable in the poem. Ubiquitous waste inspires the desire for transcendence and purification. Hence the vicissitudes of living in a desacralized urban environment are conducive to a grail quest for salvation and meaning.

There is an external and an internal wasteland which further breaks the boundaries between nature and culture in the urban environment. In *The Idea of a Christian Society*, Eliot states that "a wrong attitude towards nature implies, somewhere, a wrong attitude to God" (62). From an ecocritical perspective, the implications that sacredness is not confined to culture, but extends to nature, highlights the urban environment as a liminal amalgamation of spaces whose fluidity allows the existence of simultaneous experiences. As Terblanche explains, the poetry of T.S. Eliot "works radically against mechanism to rediscover remnants and ultimately whole patterns of organic aliveness, directly or indirectly" (188). The disparate elements of the urban environment are organically unified in a design that connects the regenerative forces of the urban wildlife with the transformative technologies of the industrialized city.

William Carlos Williams offers a special exposition of organic unity found in nature in the philosophical essay titled "Beauty and Truth," with an obvious reverent attitude to John Keats's "Ode on a Grecian Urn." He states that "all nature is but a complex arrangement" (165). He proceeds to describe his predilection for what he calls "elements," rather than "form" and remarks that nature has the distinct ability of organizing itself into a system according to a set of innate laws that permit elements to shift positions while preserving permanent truths. Nature, for William Carlos Williams, is to be observed from a distance, allowing oneself to be transported by moments of rich perception. Instead of attaching a plethora of historical, cultural or personal meanings to the poem, tantamount to T.S. Eliot, in order to reveal an environmental consciousness, William Carlos Williams lets elements of nature reveal themselves to the reader, without forcing or privileging any specific meaning. Thus, his poetry resists obscure, hyperintellectual associations and interpretations, focusing on the immediate, physical world. The material world described in his poems is unencumbered by any decorum. It is familiar, organic, ordinary and often vulgar. There is no matrix of associations involved in interpreting the poems of Williams, epiphanies usually occur from mundane images of both the animate and the inanimate world.

In his *Autobiography*, Williams refers to *The Waste Land* as "the great catastrophe to our letters" (146) that could set American poetry back twenty years by turning it into an academic exercise. William Carlos Williams would actually dedicate more than twenty years of his life to write his response to T.S. Eliot's *The Waste Land*, by employing imagist methods to construct an American epic. *Paterson* was initially conceived in 1927 but, due to the creative struggles associated with embarking on such an ambitious project, it was published years later in five volumes. The first book, which appeared in 1946 was subsequently followed by four other installments in 1948, 1949, 1951 and 1958. The distinct objectivist poetry of Williams reflects a strong belief in the power of *things* and represents a direct corollary of Ezra Pound's creed that "the proper and perfect symbol is the natural object" (109), and of the experimental art world associated with various artistic movements, from Cubism to Dadaism or Surrealism, which paved the way to the defamiliarization of mundane objects and their subsequent transformation into objects of art. "No ideas but in things" is the reoccurring dictum in *Paterson*: "Say it, no ideas but in things – /nothing but the blank faces of the houses/and cylindrical trees/bent, forked by preconception and accident – /split, furrowed, creased, mottled, stained – secret –into the body of the light!" (6). The poem itself becomes an object, language is given materiality, reality is stripped of its artificial burden and natural forces are allowed to shape the textual spaces.

In *Paterson*, nature is firmly grounded in an urban environment. The poem is set in the eponymous small town in New Jersey, neighboring Rutherford, the hometown of William Carlos Williams where he anchored himself in a suburban life and conducted his medical practice. In *Modernism and the Idea of Modernity*, Adina Ciugureanu observes that Williams surpasses his initial intention "to create an extensive and intensive image of the world, iconically represented by a city, in which opposition would end in synthesis and heterogeneity in homogeneity," embarking instead on his own quest for a deeper engagement with the real by embracing the discordant and the fragmentary (Ciugureanu 121). In a geocritical spirit, the inception of fictional Paterson is both an internal and external, private and common, an intratextual and extratextual phenomenon. Bertrand Westphal states in *Real and Fictional Spaces* that "fiction detects possibilities buried in the folds of the real" (103). If London is the "Unreal City, / Under the brown fog of a winter dawn" (*The Waste Land* 7), undone by death, Paterson is a "[b]eautiful thing" with a similar predilection for dissolution, "the whole city doomed" (*Paterson* 81). Both instances envision a dying society by illustrating the outcome of a personal and environmental collapse, thus encouraging a reconsideration of human-nature relations. William Carlos Williams unearths the public and personal history, politics and biodiversity of this town to unveil the quintessence of industrialized America and present to the reader the local effects of an environmental disaster.

Paterson represents a relevant choice of setting because of its tumultuous history on the banks of the Passaic River. The town was originally conceived in order to "lay out the prototype of a new type of American community, devoted to industry" (Roth 36). Its emergence as the epitome of the industrial city was not sustainable and mechanization took its toll on the city, incrementally turning it into a slum. The poem describes the rise and fall of the town as follows: "reversed in the mirror of its/ own squalor, debased by the divorce from learning,/ its garbage on the curbs, its legislators/ under the garbage, uninstructed, incapable of / self-instruction (*Paterson* 81). Misery and success are intermingled but what truly intertwines the real and the poetical realms is the "elemental" configuration of the environment. Felstiner regards William Carlos Williams "an environmentalist before the movement took hold" (xiv). Indeed, the local real-life experiences of in the city constitute the heart of the poem, but they are transfigured to universally express environmental understanding in the context of an extensive cultural heritage: "The province of the poem is the world. /When the sun rises, it rises in the poem/and when it sets darkness comes down/and the poem is dark" (*Paterson* 99). There is no need to search beyond nature to experience profound visionary and spiritual moments. Word pictures of American suburban imagery projected via isolated flecks of perception suffice.

The fragmentation which characterizes *The Waste Land* is also visible in *Paterson* at a textual level, since the poem incorporates heterogeneous forms, from verse to prose, from personal letters to historical excerpts from newspapers and geologic records of the city. Nevertheless, as opposed to the former, reconnection to place and nature is attained by engaging with the physical and cultural strata of the urban environment only to scrape them off. The spaces in between these sensory juxtapositions and the emotions evoked through concrete language and image work to create a new poetic universe from prosaic realities. In the words of Williams, "the poem/the most perfect rock and temple, the highest/falls, in clouds and gauzy spray, should be/so rivaled, that the poet/in disgrace, should borrow from erudition (to unslave the mind)" so that it may cast a light on "the inevitable/poor, the invisible, thrashing, breeding/debased city" simultaneously "illuminating it within himself, out of itself" (*Paterson* 80). Remnants found in the urban environment and in insignificant lives, every detail and punctuation adds to the construction of a textual terrain that scaffolds on the everyday life and brings dignity to that which would otherwise be overlooked, condemned, ostracized, thrown away. There is a symbiotic relationship between the environment and the poet whose interpretation of nature, history and politics makes something extraordinary out of commonplace horrors, in response to the state of anxiety and pain of the twentieth century.

Despite presenting the urban environment as a particular terrain for human cruelty, the two poems offer a glimmer of hope for nature and city alike to regenerate. In *Beyond Romantic Ecocriticism: Toward Urbanatural Roosting*, Ashton Nichols coins the term "urbanature" to describe the inextricable connection between man and nature. In line with the aforementioned modernist obliteration of dichotomous thinking, according to Nichols, "nature and urban life are not as distinct as human beings have long supposed" so that human and objects are "linked in a complex web of interdependent interrelatedness" (xiii). One cannot talk about a return to nature because nature can never be expelled from our existence. The very principles of biological birth and organic death compel us to describe our existence ecocentrically. The ruination of land and self in "fragments I have shored against my ruins" (*The Waste Land* 20) speaks about the process of decay that affects nature and humans alike. The same inexorable bonds tie the urban to the environment. "A man like a city and a woman like a flower," describes Williams, "who are in love" (*Paterson* 7). The divorce between the two instances is mitigated in the poetry of Williams. Nature and culture are married to each other again.

Natural processes are perceptible in the city. From the first lines, the town assumes a mythical aura, anthropomorphized in the figure of a giant who is protected by the primordial elements: "Paterson lies in the valley under the Passaic Falls/its spent waters forming the outline of his back. He/lies on his right side, head near the thunder/of the waters filling his dreams!" (Paterson 6). Throughout the poem we follow the course of the river which represents the source of vitality and myriad reverberations of the city. The fluidity of the river, which collects the by-products of culture, waste and remains, may also be correlated with human subjectivity that sifts through the stimuli of the modern city assaulting the senses. In a climate of anxiety and exhaustion, the natural phenomena associated with "giants" are confronted with what Baudelaire called "the great modern monster," *ennui* (qtd. in Nicholls 7). *Apatheia* and *aphasia* are the manifestations of an inability to establish contact with nature: "They may look at the torrent in/ their minds/and it is foreign to them. [...] the language/ is divorced from their minds" (*Paterson* 12). Thus, metaphysical poverty accompanies a language that is hardly connected to the earth. The residents "walk incommunicado" (*Paterson* 9), unable to understand or resonate with the city. The true language is that of the river, punctuated by the falls, alternating rhythms with the pulsating patterns of the wind howling, to which the poet is attuned. Essentially, the mind of the poet engages with the signifier of language not as a mediator of meaning, but as an act of creation, compassion, union. Despite having one's back turned on the sources of life, nature springs the cityscape and the mindscape to life.

The triad man-city-nature contributes to the vision of the city as a fertile ground for the explorations of intersections between the natural and the artificial. The city may be perceived as the outgrowth of nature. Perhaps an extension, a sort of second nature that has gradually replaced the original. For Eliot, a rather maladjusted extension, a space of consented physical and psychological entrapment: "Prison and palace and reverberations/ Of thunder of spring over distant mountains" (*The Waste Land* 16). For Williams, a self-reflexive extension of the self and nature, calling for unity: "Look, there lies the city! – calling with his back/ to the paltry congregation, calling the winds; /a voice calling, calling" (*Paterson* 18). The reverberations of nature intruding into the cityscape suggest "humanity's position in a nature that is always urban, an urbanity that is always natural" (Nichols 188–189). There is an organic unity in multiplicity and nature does not stop where human intervention begins. The cycles of regeneration present in nature are analogous to the creative violence of modernist poetry that engages in a dialectics of destruction and creation.

The dialectics of destruction and creation

In the construction of an objective poetry and of a poetry of objecthood, respectively, it is fascinating to observe the experimental engagement with language to highlight ecological imperatives. To some extent, nature becomes an object of aesthetic consumption. Perhaps due to his dual perspective as a doctor and poet, or by following the imperative of Ezra Pound to "Make it New," William Carlos Williams builds his poetics on the dialectics of destruction and creation which encompass both a natural and a social dimension. T.S. Eliot designs his waste land awaiting redemption in the aftermath of the destruction and desolation caused by war. Nevertheless, their *modus operandi* are different when articulating their poems. In *Wild Thought*, Claude Lévi-Strauss scrutinizes further the nature of art and the connections between nature and culture by distinguishing between the *bricoleur* and the *engineer*. When (re)creating the urban environments, the *engineer* "always seeks to open a way through and situate himself beyond the constraints that make up a given state of civilization", whereas the *bricoleur* "willingly or by necessity, remains on this side of those constraints" (23). The former operates with concepts that are transparent to reality, while the latter assembles pre-existing ideas in a game of permutations. Lévi-Strauss stresses that the signs that *bricolage* employs are "preconstrained" (22), in the sense that the *bricoleur* borrows from culture the pre-established conceptions of objects so that they already possess a sense of their meanings and limits when maneuvering them. Conversely, the *engineer* resembles a scientist who transcends culture in search of new concepts.

It might be argued that T.S. Eliot would fit the description of the *bricoleur*, engaged in an intertextual game of idiosyncratic allusions, (mis)quotations, arcane voices that create an environment of signs. William Carlos Williams would be the *engineer*, who destroys linguistic and poetic conventions and is adamant in his non-composition, choosing mindful open verse forms, even "deformed verse [...] suited to deformed morality" (*Paterson* 40), notwithstanding the risk of being considered unintellectual. His voice becomes the manifestation of the landscape. The willingness to liberate language and bring it closer to nature is overtly expressed by Williams: "unless there is /a new mind there cannot be a new /line, the old will go on / repeating itself with recurring /deadliness: without invention /nothing lies under the witch-hazel /bush, the alder does not grow" (*Paterson* 50). Williams inadvertently refers to Eliot as an engineer: "Who is it spoke of April? Some/insane engineer. There is no recurrence. /The past is dead." (*Paterson* 142). Although Eliot's reverence for the past is persistently mocked by Williams, neither Eliot nor Williams align themselves with one set of principles or philosophy, but instead bring together seemingly contradictory forces and ideas, accepting change and instability. In this regard, they are both *engineers* who work with, and, respectively, against the institutionalized association of objects for the sake of innovation.

The engineers of *The Waste Land* and *Paterson* explore the subterranean sources of the American and Anglocentric soils to create what could be referred to as rhizomatic maps of meaning. The poem as a rhizomatic map creates new paths between the word and the world by allowing poetic explorations of the urban environment. The concept of the rhizome, coined by Deleuze and Guattari in *A Thousand Plateaus*, takes from nature the structure of the rhizome, with its root stalks stretching in all directions and lack of a central structure, in order to replace the dichotomous and hierarchical thinking traditionally represented by an arborescent image. The rhizome can best be expressed in the form of a map in that "it is entirely oriented toward an experimentation in contact with the real" (Deleuze and Guattari 12). As a result, the subject "accedes to a higher unity, of ambivalence or overdetermination, in an always supplementary dimension to that of its object" (Deleuze and Guattari 6). Consequently, rhizomatic maps are associated with the principles of multiplicity, heterogeneity and bridge any cultural gaps to create new connections devoid of a single source root. The ecocritical dimension of rhizomatic maps is made clear by the fact that "[t]he rhizome itself assumes very diverse forms. [...] The rhizome includes the best and the worst: potato and couchgrass, or the weed. Animal and plant" (Deleuze and Guattari 7). As it has been argued in this essay, ecocriticism is a mosaic of meanings and interdisciplinary connections and so is modernism. Especially in the case of Williams, these rhizomatic maps are constantly evolving, to include new materials, fragments, debris that would fuel the "stirring dull roots" (*The Waste Land* 5) and "humped roots matted into the shallow soil" (*Paterson* 61).

While T.S. Eliot reflects in his poem the search for regeneration and feeds on the sap of European culture, William Carlos Williams harvests the intelligence of the American soil. This would account for the recurrent imagery of stones, particularly red stones, in both poems. Stones offer themselves to the poet to be chiseled into art since "[t]he stone lives, the flesh dies" (*Paterson* 49) but concomitantly, they may be a manifestation of nature and human nature because they are simultaneously perceived and conceived by the artist: "—his mind a red stone carved to be/ endless flight/ Love that is a stone/ endlessly in flight, / so long as stone shall last bearing/ the chisel's stroke/ ... and is lost and covered/ with ash, falls from an undermined bank" (*Paterson* 49). In *The Waste Land*, the invitation to "come in under the shadow of this red rock/ And I will show you something different" (*The Waste Land* 5) points once more to the unreality in which we are all prisoners, in Plato's vision, as well as the metaphorical veil that the poet is trying to remove. Creation for Eliot implies displacement whereas for Williams creation requires destruction and metamorphosis.

Williams turns his eyes towards the ground, towards a world of the violated and the vulnerable, of nature tainted by industrialization and greed in conjunction with the reverberations of war. Through his glorification of what is condemned, he respects the postulate of urbanature that "[t]he entire planet depends on a new willingness to see urban and nonurban spaces as equally worth saving." (Nichols xvi). Considering the very name of the town, the association between "pater," the father figure, and "son" foregrounds a filial relationship between the two instances. In *Tristes Tropiques* Claude Lévi-Strauss states that "[t]he town is perhaps even more precious than a work of art in that it stands at the meeting point of nature and artifice" (124). The city is the protagonist of the poem that, through the craftsmanship of the poet, generates an offspring, the poem. In the words of Williams, "only one man – like a city" (*Paterson* 7). The generated rhizomatic maps follow the flow of the amorphous thoughts of the modern man, his speech patterns, as well as the rhythms of the various elements of the urban environment that assault the senses.

In this respect, William Carlos Williams writes a poetry of compassion and exceptional tenderness for the marginalized. There is, however, a visceral involvement in what is represented, opposed to T.S. Eliot's impersonality that is limited to the human realm. Williams does not avert his eyes from the recurrent "beautiful thing," instantiated throughout the poem in both human and natural forms, though tarnished at times. The defilement of the land is the most visible, from the destruction of forests to the killing of animals. As in *The Waste Land*, the poet-dog unleashes the soiled voices of nature and pours them in verse: "Listen! – / the pouring water!/The dogs and trees / conspire to invent / a / Bow, wow! […]. No / poet has come, no poet has come. / – soon no one in the park but / guilty lovers and stray dogs / Unleashed! /Alone, watching the May moon above the trees" (*Paterson* 79). Perhaps his pronouncement to "Embrace the foulness/ – the being taut, balanced between/eternities" (*Paterson* 103) can be correlated with the raw and the visceral nature of his practice as a doctor. The imagery created is unsettling but it serves the aim of troubling civility and privilege.

By engaging in a dialogue with the apocalyptic vision of *The Waste Land*, *Paterson* uncovers the nature and multifarious dimensions of the "beautiful thing" that could be reestablished in the present. Part III of the poem presents a real fire that nearly destroyed Paterson and was witnessed by Williams in his teenage years. There are echoes of Carthage burning in "The Fire Sermon" (*The Waste Land* 15) in the hyperbolized obliteration of the urban environment depicted as a pastiche of artificial and natural elements: "- the whole city doomed! And/ the flames towering/ like a mouse, like/a red slipper, like/a star, a geranium" (*Paterson* 116). Even amidst destruction, or maybe due to their erasure, the elements of the city are seen as if for the first time, in a joyous rediscovery. The climactic scene is the burning of a library. This destruction may suggest that "[p]oetry, and by extension art […] should be born of experience rather than limited to stale and musty bookshelves" (Templeton 108). Nevertheless, from an ecocritical lens, the dialectics of destruction and creation unite nature and culture inasmuch as they are embedded in both natural and social phenomena. If T.S. Eliot presents a gloomy, defeatist imagery of environmental destruction and suggests the possibility of regeneration, Williams relishes the spectacle of the consuming forces. The defiant energy of the fire is celebrated as a productive site of creativity: "Poet Beats Fire at Its Own Game!" (*Paterson* 118). The two poems use images of urban environment that reveal multifarious facets of the natural and the artificial within, amassing their energies for regeneration and thus matching the destructive and creative forces of the universe with the liberating power of poetic creation.

Conclusion

The urban environments in T.S. Eliot's *The Waste Land* and William Carlos Williams's *Paterson* may be reassessed through a more recent critical lens marked by the shift from an anthropocentric perspective towards an ecocentric perspective. Nature with Eliot and Williams is an "urbanature," deeply rooted in humanity. Beyond the apocalyptic imagery of deserted wastelands barren of vitality and filled with hollow men, the linguistic intrusion of words related to the natural environment suggests the presence of absence. The poem itself becomes an act of rebuilding predicated on rhizomatic maps of meaning that create new paths between the word and the world by allowing poetic explorations of the urban environment. The use of poetic imagination suggests a deconstruction of the relationship between literature, nature and culture, a reassessment of the traditionally dichotomous categories of human and animal in order to accommodate a more genuine interaction with the material world which is submitted to a constant flux of destruction and creation. The barking of the poet-dog attempting to locate in poetic and theoretical registers the generally invisible connections between all manifestations of life would subsequently be transfigured into the countercultural howl of the Beats, as an organic continuation of visionary perspectives on urban environments.

Acknowledgement

This work has been supported by the project PROINVENT in the framework of Human Resources Development Operational Programme 2014–2020, financed from the European Social Fund under the contract number 62487/03.06.2022 POCU 993/6/13/ – Cod SMIS: 153299.

References

Buell, Frederick. *From Apocalypse to Way of Life: Environmental Crisis in the American Century*. New York: Routledge, 2005. Print.

Certeau, Michel de. *The Practice of Everyday Life*. Tr. Steven Rendall. Berkeley: University of California Press,1984. Print.

Ciugureanu, Adina. *Modernism and the Idea of Modernity*. Constanța: Ex Ponto, 2004. Print

Childs, Peter. *Modernism*. London and New York: Routledge, 2016. Print.

Deleuze, Gilles and Felix Guattari. *A Thousand Plateaus*. Tr. Brian Massumi. London and Minneapolis: University of Minnesota Press, 2005. Print.

Eliot, T.S. *The Waste Land*. Ed. Michael North. New York and London: Norton & Company, 2001. Print.

———. *The Idea of a Christian Society*. London: Faber and Faber, 1939. Print.

———. *The Selected Prose of T.S. Eliot*. Ed. Frank Kermode. New York: Harcourt, 1975. Print.

Felstiner, John. *Can Poetry Save the Earth?: A Field Guide to Nature Poems*. New Haven: Yale University Press, 2009. Print.

Fussell, Paul. *The Great War and Modern Memory*. Oxford: Oxford University Press, 2009. Print.

Glotfelty, Cheryll. *The Ecocriticism Reader: Landmarks in Literary Ecology*. Athens and London: The University of Georgia Press, 1995. Print.

Harrison Pogue, Robert. *Forests: The Shadow of Civilization*. Chicago: University of Chicago Press, 1993. Print.

Heise, Ursula K. "The Hitchhiker's Guide to Ecocriticism." *PMLA*, vol. 121, no. 2, 2006, pp. 503–16. Print.

Lévi-Strauss, Claude. *Wild Thought*. Tr. Jeffrey Mehlman and John Leavitt. Chicago and London: University of Chicago Press, 2021. Print.

———. *Tristes Tropiques*. Tr. John Weightman and Doreen Weightman. New York and London: Penguin Classics, 2012. Print.

Nicholls, Peter. *Modernisms: A Literary Guide*. New York: Palgrave, 2009. Print.

Nichols, Ashton. *Beyond Romantic Ecocriticism: Toward Urbanatural Roosting*. New York: Palgrave, 2011. Print.

Philips, Dana. *The Truth of Ecology: Nature, Culture, and Literature in America*. Oxford and New York: Oxford University Press, 2003. Print.

Pound, Ezra. "Retrospect." *Prose Keys to Modern Poetry*. Ed. Karl Shapiro. New York: Harper and Row, 1962. 104–112. Print.

Roth, Leland M. *America Builds: Source Documents in American Architecture and Planning*. New York: Routledge, 2018. Print.

Sullivan, Heather I. "Dirt Theory and Material Ecocriticism." *ISLE: Interdisciplinary Studies in Literature and Environment*, vol. 19, no. 3, 2012, pp. 515–531. Print.

Templeton, Erin E. "Paterson: an epic in four or five or six parts." *The Cambridge Companion to William Carlos Williams*. Ed. Christopher McGowan. New York: Cambridge University Press, 2016. 101–114. Print.

Terblanche, Etienne. *T.S. Eliot, Poetry and Earth: The Name of the Lotos Rose*. London and New York: Lexington Books, 2016. Print.

Tobias, Michael Charles, Jane Gray Morrison. *On the Nature of Ecological Paradox*. New York: Springer, 2021. Print.

Williams, William Carlos. *Paterson*. New York: Penguin Books, 1983. Print.

———. *The Autobiography of William Carlos Williams*. New York: New Directions Books, 1967. Print.

———. "Beauty and Truth." *The Embodiment of Knowledge*. Ed. Ron Loewinsohn. New York: New Directions, 1974. 159–169. Print.

Westphal, Bertrand. *Geocriticism: Real and Fictional Spaces*. Tr. Robert T. Tally. New York: Palgrave Macmillan, 2011. Print.

Williams, Raymond. *Keywords: A Vocabulary of Culture and Society*. Oxford: Oxford University Press, 2015. Print.

Oana-Celia Gheorghiu

Re-Imagining *The Waste Land*: Infertility, Barrenness and Ecocatastrophe in Margaret Atwood's *The Handmaid's Tale*

Abstract: Using the "chessboard" of T. S. Eliot's *The Waste Land* as an intertextual reference point, this chapter re-reads Margaret Atwood's *The Handmaid's Tale* with a view to highlighting the *wasteland* of womanhood suppression and women's rights abolishment following an environmental and political catastrophe.

Introduction: The Queen's Gambit

Aside from the writing/reading exchange, "the games of literature" (Bruss 153) often entail a reading/rewriting communicative situation, combining bits and pieces of textualized images to create a polymorphous whole, generating its own polyphony. "Much like a game of chess, [...] literary games exist wherever praxis and strategy provide the principal meaning of the work" (Bruss 153). The end result, though "a literary puzzle" (Praisler 2017: 184), is usually more cohesive than it would have been had the writers taken the advice of Tristan Tzara at face value, and would have:

> take[n] a newspaper; take[n] a pair of scissors; chose[n] an article as long as you are planning to make your poem; cut out the article; then cut out each of the words that make up this article and put them in a bag; shake[n] the bag; then take[n] out the scraps one after the other, and cop[ied] conscientiously in the order in which they left the bag. (Tzara 107)

The poem that I intend to bring to the game table, and to use as an imperturbable King in this game of chess is T. S. Eliot's *The Waste Land*, whose 433 lines seem, at times, to have been arranged following the Dada principle above. Considered by critics the highest achievement of Modernist poetry – with some overbidding, regarding it as *the* greatest twentieth-century poem – *The Waste Land* is the perfect example of a puzzle, which has not yet managed to be unequivocally glued together, and which is therefore still awaiting creative or insightful solutions for sorting out the (apparently) intended Dadaist disorder. Telling in this respect is Virginia Woolf's first reaction to the poem:

> Eliot dined last Sunday & read his poem. He sang it & chanted it, rhythmed it. It has great beauty & force of phrase: symmetry; & tensity. What connects it together, I'm not

so sure. But he read till he had to rush… & discussion thus was curtailed. One was left, however, with some strong emotion. *The Waste Land*, it is called; & Mary Hutch, who has heard it more quietly, interprets it to be Tom's autobiography – a melancholy one. (2, 178)

Equally revealing are the inexplicit comments of contemporary critics, who describe it as "disconnected, confuse" (Seldes in North 138) or as "a collection of flashes," having "no effect of heterogeneity" (*Times Literary Supplement* in North 137).

Eliot's poem – with its "heap of broken images" (I, 22), resulting from the central theme and its representation: the world's barrenness in the wake of an apocalyptic Great War which left indelible marks on human consciousness, is paired in my chess game with a Queen that is sixty-something years its junior. According to the rules of the game, she has greater liberty of movement, vertically, horizontally, and diagonally, through her creating a projection into an unknown future, through speculations on the horrendousness of the things to come.

Again, to no one's surprise, since it is in the title too, the reference is to Margaret Atwood's *The Handmaid's Tale*, a novel that is far from being as experimental and obscure(d) as Eliot's poem, but which provides, nonetheless, some of that Eliotesque heap, for the reader to arrange and make sense of it herself. It remains to be seen whether this reader can clean and arrange the pieces in an orderly fashion as if she diligently cleaned her personal space or domestic environment.

Alternatively, it remains to be seen whether she complicates things further, weaving and unraveling additional threads like an unruly Penelope, retelling both the poem and the novel in a manner that defies the norms of literary criticism, and nears, arrogantly perhaps, those of postmodern rewriting based on the revisitation of inlaid intertextuality.

A reversal of stereotypical gender roles

I have implied a preponderantly feminine readership for this novel because it outlines a fictional feminine universe, full of rooms "where women come and go / [t]alking of Michelangelo,"[20] although it is men who keep them "in check".

---

20 Allusion to Eliot's refrain in *The Love Song of J. Alfred Prufrock* (1915). Significant along the lines set here is the oblique reference to women's bodily fecundity, in sharp contrast with men's artistic fertility. The dichotomic pair is carried through and developed by T. S. Eliot in his *The Waste Land* (1922), then further exploited, subversively, by Margaret Atwood in *The Handmaid's Tale* (1985).

Incidentally, the author of *The Waste Land* is a man, whilst that of *The Handmaid's Tale* is a woman. The deafening polyphony of the former is, accordingly, mostly masculine, while the monological discourse of the latter is explicitly feminine, which produces an interesting reversal of stereotyped gender roles, since poetry is more likely to be associated with feminine sensitivity, while the novel has been perceived, traditionally, as a male territory until not so long ago (see the literary canon), with a few notable exceptions. The title of the poem suggests complete and irreversible sterility, while the title of the novel, alluding to a female character in the Old Testament who is impregnated as an ancient surrogate for her barren mistress, emphasizes the possibility of redemption through feminine fruitfulness and the ability to bear children. "He who was living is now dead" (V, 328), but for *her*, there is still a chance for the Lord to open. In this respect, although Atwood's novel truly is one of the grimmest dystopias ever imagined, although its represented oppressiveness is unbearable, and although its plot is terrifying due to announcing a future reality, or at least a darker version of the present situation, it leaves a ray of hope (particularly for the faithful) penetrate through the unimaginable nightmare that is the women's fate in Gilead.

After death, renaissance...

Similarly, limited optimism may also be read into Eliot's poem eventually, but not until the very end. In the poem's opening lines, however, "April is the cruelest month" since it is "breeding lilacs out of the dead land" (I, 1–2), is neither revival nor renewal. It just stirs painful memories of a living world that is now dying, having reached the end of a cycle – a recurrent theme in modernist poetic discourse.[21] There is no water, as the poem stresses with its eighteen mentions of this absence, therefore there is no life. Dryness has taken over the land turned to waste, and only death seems to have a chance at breeding:

> That corpse you planted last year in your garden
> Has it begun to sprout? Will it bloom this year?
> (I, 71–72)

The theme is carried through across literary genres and successive generations of literary writers, which indicates yet another cycle, at whose core lies the recycling process itself – brought forth via mythical textures and metatextual strategies. Rebirth and Renaissance are rooted in the awareness of chaos and meaninglessness,

---

21 William Butler Yeats's mystic, "occult" poems come to mind in this respect.

criticize the status quo, and propose change, but come at the price of sacrifice and therefore necessarily contain their opposite: death. Eliot and his great contemporary, James Joyce, in *Ulysses*, tackle this aspect of life that is taking its sap from the dead ("Dead breaths I living breathe, tread dead dust, devour a urinous offal from all dead," (Joyce 46)) and both refer, indirectly, to the drowned god, a fertility symbol mentioned by Jesse Weston in *Ritual and Romance*:

> [...] whose effigy was thrown into the water as a symbol of the death of the fruitful powers of nature but which was taken out of the water as a symbol of the revivified god. (Brooks in ed. North 194)

Both *Ulysses* and *The Waste Land* quote from Ariel's song in *The Tempest*: "Full fathom five thy father lies" (Joyce 46) and "those were pearls that were his eyes" (Eliot, II, 125), shaping a reusable intertext and advancing an idea which is also obliquely woven into Margaret Atwood's dystopia (with its end-of-the-world, death-of-God philosophy, and its message of combating decay by infusions of new life/mentality).

Awaiting "re-naissance" in *The Waste Land* shows (urban) civilization as Eliot knew it. In the wake of the Great War, its demise takes material form in the gloomy image of the "unreal city under brown fog" (I, 60–61), its darkness amplified by the environmental disasters left behind – "roots that clutch" (I, 19), "stony rubbish" (I, 20) and "dead trees" (I, 23).

The desperate recognition of societal collapse and the acknowledgment of ecological degradation permeate the entire work and support the violent chess played by the world, strategically placed at its center. Divided between 1914 and 1918, though won by one party, historically and politically speaking, the game ended in stalemate, for humankind lost its vital resources and is now dying, unable to find shelter, relief, and the will or the power to be reborn. The spirit of *The Waste Land* reproduces the Zeitgeist, being familiar in the period between the two world wars: the disgust and despair with everything that haunted the survivors of what Gertrude Stein called the lost generation. "The grim war behind them served as a harrowing commentary on the empty lives and venal pursuits they saw all around them" (Bloom 20). The nursery rhyme "London Bridge is falling down, falling down, falling down" (IV, 426) informs, in childish language with adult overtones, that the disintegration of civilization has not stopped at the ceasefire, and that all the King can do now, in a retelling of the legend of the Fisher King (acknowledged by Eliot in one of the few unambiguous explanatory notes he provides to the poem), is "sit upon the shore and fish with the arid plains behind [him]" (IV, 423–424), as sterile as his lands. Sterility here is synonymous with death. Checkmate, but the will for a return match/revenge is present – "Hieronimo is mad againe"[22] (V, 431), and so is hope, at the very end of the poem, or at least this is what the final mantra in Sanskrit seems to suggest. In *Reading The Waste Land from the Bottom Up*, a recent study intended as "a manual providing insight into the poem's individual components," Allyson Booth asserts that the sacred ending lines of the Upanishads can be construed in line with "nearly every literary source Eliot cites in the poem [as] a possible recasting of the whole poem: burial rite, revenge play, river song, fertility ritual, prophecy, and prayer are just a few of the available reconstructions" (252).

Nature and womanhood… victimized

Just like Eliot's poem, the novel I am using as the Queen in my game is imbued with both social and political awareness and eco-consciousness. The latter should not be read in the scientific sense of the term, as "a reflection of […] man's relationships with nature, which expresses the axiological position of the subject in relation to the natural world" or as "a complex mental education, which includes cognitive, regulatory, emotional, ethical, and other aspects," (Popov 280) but rather in relation to a feminine/feminist eco-consciousness observant of the textual politics that brings together nature and womanhood, and considers woman as Earth/ Gaia/ mother of all humankind. However, women are portrayed in *The Handmaid's Tale* as victims of the abolishment of their rights after a series of environmental and political catastrophes, scarcely described in flashbacks, rather than fully outlined as life generators, although this latter stance is essential in Atwood's both breaking down reality and re-making a fictional universe to interrogate it. They choose a silent and humiliating survival that may find echoes in the quiet resistance of nature in the face of the atrocities that the urban (male) civilization brings upon it.

The Handmaid's Tale can be regarded as a literary rebuttal of patriarchy, which was defined as

> a political-social system that insists that males are inherently dominating, superior to everything and everyone deemed weak, especially females, and endowed with the right to dominate and rule over the weak and to maintain that dominance through various forms of psychological terrorism and violence. (hooks 1)

With Margaret Atwood, in this novel, as well as in others, the victim takes the shape of the central character, "whose life story is forwarded via the narrating *I*,

---

22 The line alludes to one of the most famous Elizabethan revenge tragedies, Thomas Kyd's *The Spanish Tragedy, or Hieronimo Is Mad Again* (written between 1582 and 1592).

while survival breathes through each and every detail, forming the intricate web of the events narrated" (Praisler 2014: 94).

To preserve their womanhood at least, after they have lost their identity and freedom, some victims, like Offred, must become mothers, whose children will be taken away from them, but mothers nonetheless; they must fulfill their natural, biological function of giving birth and of perpetuating the endangered human species.

Not all enslaved women endure the confinement and rape that should lead to this incomplete maternity outcome. The narrated Offred, introduced before the narrating Offred, experiences conflicting states of mind, which may be decoded from analyzing the way in which she reacts to texts and contexts, and which range from rebellion/activity – in scrawling the mock-Latin prompt, "*Nolite te bastardes carborundorum*" (Don't let the bastards grind you down!) on the walls of her "prison" – to submission/passivity – in being overwhelmed by a death wish which resonates of the epigraph in Latin and Greek borrowed by Eliot from *The Satyricon* to open his poem with:

> Nam Sibyllam quidem Cumis ego ipse oculis meis vidiin ampulla pendere, et cum illi pueri dicerent: Σίβυλλα τί Θέλεις; respondebat illa: ἀπο Θανεῖν Θέλω.
>
> (For I once saw with my own eyes the Cumean Sibyl hanging in a jar, and when the boys asked her, 'Sibyl, what do you want?' she answered, 'I wish I want to die'). (Eliot 3)

If, as Booth rightly notices, the prophetess "presides over *The Waste Land* from her bottle – near death, wishing for death, as helpless as if she were dead, but unable to die" (35) – the woman caged in an environment twistedly remindful of a woman's traditional place, the household, eventually has her wish fulfilled, committing suicide. She thus leaves room (both connotatively and denotatively) for Offred to weave her story at *Night(s)* and to live a life of domestic abuse during the day.

Breaking the rules of the game

The dangerous, underground game of "telling" that Offred plays is contrasted with a game of patience and anticipation, Scrabble, which she is exceptionally allowed to play (be it only for the entertainment of the bored Commander she is assigned to). Requiring a smaller degree of abstraction than chess (and therefore more appropriate for a woman), it uses words but abuses language, since neither meaning nor context are sought after. Rules are present in this game as much as they are in chess, but it is the male element that imposes them on the Handmaid while breaking his and his peers' own rules. Offred is invited to break these rules with him, but her role is precisely to abide by new rules. As for abiding by no rules, being subject to no patriarchal authority (as is the case of her secret recordings) it is out of the question. This makes her endeavor to communicate against all odds all the more daring, all the more noteworthy, and may be interpreted as defiance, as refusal to accept any kind of restriction or violation. In an analysis of Eliot's only section of *The Waste Land* that is not elemental (the other four referring to earth, fire, water, and thunder), namely *A Game of Chess*, American New Critic Cleanth Brooks accounts for T. S. Eliot's source of inspiration, Thomas Middleton's play *Women Beware Women* (1657), acknowledged by the poet himself in the note which explains line 138:

> In the play, the game is used as a device to keep the widow occupied while her daughter-in-law is being seduced. The seduction amounts to almost a rape, and, in a double entendre, the rape is actually described in terms of the game. We have one more connection with the Philomela symbol, therefore. The abstract game is used in the contemporary waste land, as in the play, to cover up a rape and is a description of the rape itself. (Brooks in ed. North 195)

The game of Scrabble into which the Commander forces the woman in his power can easily be read along similar lines. Even the character–narrator makes the same observation:

> What had I been expecting, behind that closed door, the first time? Something unspeakable, down on all fours perhaps, perversions, whips, mutilations? At the very least some minor sexual manipulation, some bygone peccadillo now denied him, prohibited by law and punishable by amputation. To be asked to play Scrabble, instead, as if we were an old married couple or two children, seemed kinky in the extreme, a violation in its own way. (Atwood 163)

The game, together with the allusion to empowerment through language, remains, in the Atwoodian text, a symbol of imposition, control, and, in the end, meaninglessness. It is a game in which the woman always loses, even when she scores more points than her male opponent. Words cannot make sense on their own, and the table, with its disparate letters and mismatched words heading in all directions, is incapable of creating meaning (unless one takes the Dada strategy (of fighting back fragmentation with hyperbolized fragmentation) a step further and uses the words to create an unintelligible poem). It is Offred who remarks that "context is all" (Atwood 154) somehow supporting the notion that the dissolution of the real cannot but be reflected in incoherent language, generating unbelievable representations. The specific reference, however, is to the Gileadean context, to its violence and its absurdity. In Gilead, real access to language is denied to women; actual communication is prohibited; reading and writing bring about the death penalty. An enforced game of Scrabble is almost a form of rape, which is added to the other (at least) two rapes inflicted on the feminine other – the rape of women, attempting to restore their fertility, but only seeding trauma, and that of the planet herself, which has eventually led to sterility.

The alternative for women in Gilead is illiteracy and domesticity. "They're very clean-minded, these days" (260) says Moira, Offred's friend who has escaped the handmaid's training center just to end up a prostitute at the Jezebel's, one of the most hypocritical "institutions" of the theocratic Gilead. The irony of the double meaning is more than obvious, especially when uttered in a brothel, by a prostitute at the disposal of the presumably pious ruling class. But the character does not refer (at least, not directly) to the cleanness of the mind and conscience, but to the practical activity of cleaning, which, under the new order, is the task of *less women* if it takes place in a domestic environment, and that of the *least women (Unwomen)* in the darker context of the wasteland that lies beyond the cities. The former, housekeepers and cooks, are also charged, where the case may be, with the task of another form of surrogacy (that of taking care of the children for the purely decorative Wives). The Marthas, as they are called, are older, and infertile, and it is also alluded that they may belong to racial minorities (Black or Latinas). With their lower status and inferior ranking in the social stratification of Gilead, it is the Marthas who best embody stereotypical Women in a male-controlled society; they are the "mothers scrubbing the floor" of this hyperbolized patriarchate where "'Home is the girl's prison, / The woman's workhouse" (Fanthorpe 1988). The latter, punished for being different or for having made life choices which either inconvenience or go against the mainstream, pay the heavy price of being set the most menial tasks, and are generally considered disposable.

From an environmentalist–cum–social perspective, *The Handmaid's Tale* displays some connections with *The Waste Land*, whether Atwood considered them or not. The premise of both texts is sterility, which affects not only natality but the natural environment also – as in the legend of the Fisher King. Atwood has her caricatural academic, Peixoto, scientifically explain, in the coda to the novel, *Historical Notes*, the severely reduced birth rates and the presence of some dreaded territories named Colonies, full of toxic waste, which the lowest ranks of the population, the Unwomen (including the Gender traitors, i.e., homosexuals), are sent to clean, perhaps to be, in this way, (re)assigned a woman's traditionally subservient role, one that they lost upon having the negative prefix *un-* added to their gender:

> Still-births, miscarriages, and genetic deformities were widespread and on the increase, and this trend has been linked to the various nuclear-plan accidents, shutdowns, and

incidents of sabotage that characterized the period, as well as to leakages from chemical and biological warfare stockpiles and toxic waste disposal sites, of which there were many thousands, both legal and illegal – in some instances these materials were simply dumped into the sewage system – and to the uncontrolled use of chemical insecticides, herbicides and other sprays. (Atwood 317)

Inside the novel proper, however, there is no scientific explanation for the existence of such *waste lands*, but there is a horrific, albeit short description of them, which capitalizes on death imagery: "just bodies, after a battle ... left around longer, they get rottener," "your nose falls off and your skin pulls away like rubber gloves." "The toxic dumps and the radiation spills" are the fundamentalist regime's slow-death punishment for people they want to get rid of: "old women [...] and Handmaids who've screwed up their three chances, and incorrigibles like me. Discards, all of us. They're sterile, of course" (260).

Conclusion: Checkmate? Or is there any hope?

The open endings of both texts brought together, if for no other reason, at least for their somber vision of a world left in disarray after a political and ecological catastrophe, leave a thin ray of hope for the reader to shake off the sensation of impending doom. Eliot's "Shantih" (IV, 433) is "repeated as here, a formal ending to an Upanishad; The Peace which passeth understanding is our equivalent to this word" (Eliot in North 26). The "translation" provided is a quotation from Philippians 4:7 and refers to spiritual enlightenment and religious salvation. By contrast, Atwood implies a possible salvation not by, but *from* religion in the last sentence of the novel, which sends Offred "into the darkness within, or else the light" (307). Could darkness and light symbolize cyclicity and, with that, a perpetual new beginning, like Eliot's cruelest April? Could the light, on the other hand, signify redemption through death (an ironically Christian stepping into the light), in which case, provided that she was not pregnant (the novel does not offer certainty in this respect), the world's sterility will be healed by her disappearance, as the Fisher King's world would flourish again upon his death? One cannot know for sure, nor is it of any use for one to hypothesize what happens *after fiction* or *after poetry*. It may be, notwithstanding, useful to pay attention to what happens at the same time with the production of the said prose or poetry in the real world, to look for context, and to read the literary writings as warnings against the destruction of the natural world, in turn engendering an untimely end to human civilization. Whether one conceptualizes this destruction either as a war or as a game of chess, it would be but a disastrous win.

References

Atwood, Margaret. *The Handmaid's Tale*. London: Vintage, 2010. Print.

Bloom, Harold. *Bloom's Guides. T. S. Eliot's The Waste Land*. New York: Infobase Publishing, 2007. Print.

Booth Allyson. *Reading The Waste Land from the Bottom Up*. New York: Palgrave Macmillan. 2015. Print.

Bruss, Elizabeth W. "The Game of Literature and Some Literary Games". *New Literary History*, vol. 9, no. 1, 1977, 153–172. Web.

Eliot, Thomas Stearns. *The Waste Land. T. S. Eliot. A Norton Critical Edition*. Ed. Michael North. New York and London: Norton & Co., 2001, 3–29. Print.

Fanthorpe, U. A. "Mother Scrubbing the Floor". *Atlanta Review*. Spring/Summer 1988. Web.

hooks, bell. *Understanding Patriarchy*. Louisville Anarchist Federation, 2010. Web.

Joyce, James. *Ulysses*. London: Wordsworth, 2010. Print.

North, Michael (ed.). *The Waste Land. T. S. Eliot. A Norton Critical Edition*. New York and London: Norton & Co., 2001. Print.

Panov, V. I. "Ecological Thinking, Consciousness, Responsibility". *Procedia – Social and Behavioral Sciences*, vol. 86, 2013, 379–383. Web.

Praisler, Michaela. "An Exercise in Representing Memory: Margaret Atwood's *Surfacing*". *Comunicare Interculturală și Literatură*, vol. 2, 2014, 94–101. Web.

Praisler, Michaela. "Hi(s)story Gone Wrong. Martin Amis on the Holocaust in *Time's Arrow*". *Cultural Intertexts*, vol. 7, 2017, 183–195. Web.

Tzara, Tristan. "To Make a Dadaist Poem". In Pericles Lewis (ed.) *Cambridge Introduction to Modernism*. Cambridge University Press, 2007, 95–126. Print.

Woolf, Virginia. *The Diary of Virginia Woolf. Volume Two: 1920–1924*. New York and London: Harcourt Brace Jovanovich, 1978. Print.

Michaela Praisler

"Is this the house you want to live in?" Fictional Warnings from a Feminine Eco Consciousness

Abstract: Highlighting environmental wrongs and women's rights, Margaret Atwood's *The Handmaid's Tale* also exposes puzzling times and the literary puzzle through clues woven into its fabric and through the closing paratext. In touring her house of fiction, this chapter also considers her associated non-fictional interventions – interviews, lectures and speeches.

Introduction

North-American literature has a centennial tradition of approaching nature along the lines of the already well-known list of dichotomies – man/woman, urban/rural (pastoral), active/passive –, opposing terms which many scholarly voices are struggling to blur, even to interchange. With the passing of time, technological advancement and general human progress have inevitably influenced fiction, which has displaced its early praise of nature to a new paradigm, in which nature is either destroyed by or destroys man with a vengeance.

The rise of dystopian, catastrophic literature has entailed the rise of eco-consciousness and eco-criticism, with the corresponding environmental activism – now having become the norm rather than the exception against the background of climate and milieu related problems multiplying endlessly. Among the many branches of eco-criticism that ensued during the last three decades of the twentieth century and that are starting points for contemporary debates, particularly persistent is that which brings together environmental wrongs and women's rights: ecofeminism, which calls attention to the "connection between the exploitation and degradation of the natural world and the subordination and oppression of women." (Thorpe 2016)

It is the conceptualisation in the Western intellectual tradition of women as part of nature which, in ecofeminists' view, has resulted in "devaluing whatever is associated with women, emotion, animals, nature, and the body, while simultaneously elevating in value those things associated with men, reason, humans, culture, and the mind." (Gaard 5) It is what has contributed to justifying political agendas and what has led to men being in a position to affect environmental decision making, while women form the majority of the people who are at the mercy of those decisions. (Mellor 2003) Aside from the ideological aspects related to women and the environment, physical elements have also been taken into consideration. Scientifically documented explorations of the status quo have revealed that "toxic pesticides, chemical wastes, acid rain, radiation, and other pollutants take their first toll on women, women's reproductive systems, and children." (Gaard 5) Ideology and biology activate the constant questioning of the existing condition, and underpin the discussion on where *she* is (Cixous 1975) in the texture of cultural spaces, but more importantly on how she is, in relation to environmental destruction v preservation.

Stepping outside "the desert of the real" (Zizek 2002) and entering the realm of representation, one finds similar, though perhaps more oblique, preoccupations with gender in association with social and ecological states of affairs, as well as to religion and politics, historical circumstance, and the cataclysms lying ahead due to violently imposed hierarchies. Women's writing, in particular, even if not always openly confessing to a feminist penchant or sustaining related environmental causes, inscribes itself within broader humanistic questions regarding culture, values and ethics, while still being read through lens which highlight fossilised standpoints whereby women are either objectified or brought close to nature (in need of subjugation, exploitable for patriarchal gains).

Margaret Atwood's eco blueprint and house of fiction

Struggling to get the ecofeminist message across and expose the menace of correlated complacency, Margaret Atwood's literary work is supported by paratextual interventions by means of which the writer discloses her politics and expounds concerns like creating probable futures, building imaginable dystopias and interrogating global crises. In a talk on "Fiction, the Future and the Environment", during an interview with Eveline van der Ham (Nexus Instituut, *YouTube* 2012), she explains her strategy and her topic of predilection:

> I write about something called the future, which is a wonderful thing to write about because nobody can fact check it. But I try to base my futures on realities that are with us today. [...] And everything that I put in has a basis in reality, something we're doing now, something we've already done, something we're thinking of doing. I write about probable futures... possible and probable. (Nexus)

Her fiction emerges therefore as a continuation of the real, as a plausible destination we are all heading towards. Far from being a tourist attraction, once visited, it is impossible to escape from. Politically, the trap it sets is currently in the making. Environmentally, its bleakness is the result of the decisions made these days. And yet, it is preferable to no destination at all. Assessing our interventions in nature, Atwood raises awareness of the pluses and minuses of science and technology, indicating how imperative it is to maintain the balance, warning against contributing to destroying modern civilisation and bringing nearer the end of the world.

> Our ability to discover and invent has allowed us to open the biggest Pandora's box in the world, which is the ability to create new biological species. That can be a good thing (better potatoes) or it can be a bad thing (more destructive microbes). […] The most important thing for the world as a whole is to maintain the oxygen level that allows us to live on land. […] And then we can talk about the rest. (Nexus)

In the circumstances of today's Covid-19 pandemic and its aftermath, her words resonate more profoundly, somehow showing that the future envisaged is already here. And fact checked. However, with Margaret Atwood, the mutual contamination of fiction and reality is not a mere philosophical exercise in deciphering the interchangeability of truth and speculation. It acquires political overtones. Inside and outside the actual literary text, the writer's voice is frequently heard on neuralgic subjects like pollution, global warming, waste disposal, deforestation or ocean acidification – all of which require active and rapid governmental involvement, by collective exertion of pressure if necessary. Her stance, as formulated in a 2019 interview with *Krishnan Guru-Murthy* for Channel 4 News, on *Facebook*, is one that acknowledges the power of the younger generation (whose own posterity is at stake) and the small but efficient role of public figures (who are often in the spotlight).

> The people who are really changing things and moving the needle right now I would say are the Extinction Rebellion teenagers who are putting a bit of fright into politicians because pretty soon they're going to be able to vote. So […] 'get your head out of the sand, there is a climate crisis'. So I think that's the kind of thing that actually changes things. As a writer, you can do… or as a writer, citizen and old person like me you can do a certain amount of cheerleading. (Channel 4)

In Atwood's case, overt cheerleading in the various media is doubled by covert representations of contemporaneity in her writing. Her house of fiction is deliberately built on contorted foundations and shows unstable scaffolding, in an attempt at prefiguring the architecture of the world predicted. The construction plans, a dystopian design, frame undersized rooms, narrow corridors and dungeon like basements with the potential to accommodate fear, tyranny, enslavement, ecological disaster – bad tenants who have already been tolerated for too long by numerous unassertive or unsuspecting householders blinded by professed democracy and apparent well being.

> What the dystopia as a form does, it creates a blueprint and says to people is this the house you want to live in? And, if not, what house do you want to live in? So if you don't want to live in this world, where you seem to be heading, possibly you might consider another kind of world, that you do want to live in, and how to create that other kind of world. (Channel 4)

Creating worlds in fiction is a demanding enterprise, with a wide variety of readings and end results. In general, however, the choices tend to gravitate towards utopias or dystopias. If utopias are outlined, readers are offered ways out of the material, everyday life, without the added critique of the world that is. If, on the other hand, dystopias are the preferred alternative, the reversals and/or distorsions of these imagined 'bad places' are a means of examining the current trends, as well as the societal, political and spiritual norms or systems in place. Atwood opts for combining "the imagined perfect society and its opposite" in her 'ustopia', since "each contains a latent version of the other." (*The Guardian*) In so doing, she points to the presumed normality and the danger within, tracing potential trajectories and future metamorphoses. Margaret Atwood's ustopia (juːesˈtəʊpiə; ʌsˈtəʊpiə) seems to be rooted in the American space, but encompasses us all, with local realities conducing to global contagion by the stratagems of a world power/leader. Significantly, it exhibits inertia, overlapping the future with the past (as she propounds in a 2020 interview with Janira Gomez for France 24 on *YouTube*).

> For a while, in the twentieth century, we thought we were going in the direction of more equality, more freedom, but then things started to turn around and go the other way. [...] Right now, in the United States, many individual states are restricting rights for women in a way that we haven't seen since the 1950s. That's their Golden Age. They would like to go back there. (France 24)

"They," according to the novelist, are usually men who come to power and establish authoritarian governments "at times of fear and social chaos," who push back women regardless of ideological or religious orientation, and who pretend there is no climate crisis. Their electoral rhetoric is geared towards getting "rid of the chaos" and things being "very tidy," which makes their campaigns successful, yet hides the eventuality of paying the heavy price of killing, supressing or disempowering various groups, of silencing individual and concerted voices – many of which speak up on ecological issues – and, by consequence, of remaining oblivious to all things related to the social and natural environment. (France 24)

Under focus: *The Handmaid's Tale*

These and other taboos (like xenophobia or reproductive health) are openly tackled in Margaret Atwood's fiction which, as suggested, recycles reality, looks into myth making and myth breaking, appends her story to history, and adds a feminine-feminist twist to the male dominated literary canon. In the particular case of *The Handmaid's Tale*, she puts "nothing into the book that did not have a precedent in history or in present day practices," (France 24) selecting three relevant sources:

> Puritan seventeenth century America, which is pretty interesting… not fun for women… in fact a lot of women who got captured by indiginous people wanted to stay with them because they were having more fun; […] the history of the speculative fiction dystopia (most of which had been written by men, so I wandered what that would look like if you turned it around at that time); […] history and the world that we are living in. (France 24)

Dedicated to Mary Webster[23] and Perry Miller[24], the novel discloses its author's interest in early American intellectual history, formulates a caveat concerning the fusion of fact and fiction, and foregrounds the denigration and victimisation of women. These clues into the novel's content of ideas are supported further by three mottoes which foreshadow Atwood's fictional version of puzzling times in her literary puzzle, and which cover Christianity, Islam and bad politics: the first is from *Genesis, 30:1-3* and tells the story of barren Rachel who, jealous of the fertility of her sister-concubine, beseeches her husband, Jacob, to "go in unto" Bilhah, the handmaid, that "she shall bear upon [her] knees" and give her children; the second is from Jonathan Swift's dark satirical essay "A Modest Proposal (for Preventing the Children of the Poor from Being a Burden to their Parents or Country, and for Making them Beneficial to the Publick)," which addresses the perversion of human values in periods of oppression, class conflict, economic inequality, poverty, and famine; the third is a Sufi proverb that speaks about the power of enforcement/ prohibition for prohibition's sake – denying

---

23 In 1683, in colonial New England (Hadley, Massachusetts), Mary Webster, commonly known as Half-Hanged Mary (also the title of a poem by Margaret Atwood), was accused of witchcraft. Found not guilty by a court in Boston, she was later left hanging by neighbours, only to be found still alive the next day. Atwood says that it is possible that she may be related to her.
24 Perry Miller was a professor of History of Ideas at Harvard University, specialising in the age and worldview of Puritanism. Atwood studied under him at Harvard Radcliffe Institute.

no actual desire, being absurd therefore. The three corresponding images and themes are entwined into the novel's fabric and give potency to its overall message. The apocalyptic, dystopian narrative imagined by Margaret Atwood in *The Handmaid's Tale* features the ritual of begetting children – imported from the Bible and read literally –, the practice of children used and abused as currency or "fodder," an establishment based on the adults' unnecessary display of administrative and military might.

Atwood's rocky Gilead, giving testimony to the aftermath of the world's having plunged in disarray, is the result of a hostile takeover of the United States by the Sons of Jacob, a group of religious fundamentalists who impose a martial, male-only dictatorship and who give the women found in the wrong three equally bad options – prostitution in brothels (the Jezebels), slow death by toxic pollution (the Unwomen) or consented rape and the questionable honour of becoming "sacred vessels, ambulatory chalices" (Atwood 146) (the Handmaids). Their agenda is rooted in a fertility crisis, not entirely fictional, since it was inspired from the realities of the 1980s, when the novel was published. Sterility, infertility, miscarriages, deformities or infant deaths lead to the implementation of a politics of repopulation. It is women's fault in Gilead's official discourse: "He's said a forbidden word. *Sterile*. There is no such thing as a sterile man any more, not officially. There are only women who are fruitful and women who are barren, that's the law." (Atwood 71) Can biology be regulated by law? Of course not, and this is one of the many possible avenues of opening the debate on silencing women and on men's dominant positions. The might is right to the extent that it bans words and enforces its own authoritative truths.

Much more than just patriarchal, the Gileadean society is phallocentric and phallologocentric (with the latter remaining obvious even in the "mansplaining" coda of the novel). Women fall prey to abuse gradually, unknowingly, not without being warned, but ignoring the warnings. The tale, Chaucerian by name and by rough obscenity, far from signifying nothing, but screaming warnings against a conceivable future reality from its fictional vantage point, is told by an idiot narrator who would not listen to the sound and the fury surrounding her, and would eventually find herself imprisoned in a *gynaeceum*, to be heard no more – June/ Offred ('of Fred' Waterford). This future (actual past, since history repeats itself), perceived as fabrication or disregarded altogether before Gilead, is allowed to materialise.

> Nothing changes instantaneously: in a gradual heating bathtub you'd be boiled to death before you knew it. There were stories in the newspapers, of course, corpses in ditches or the woods, bludgeoned to death or mutilated, interfered with, as they used to say, but they were about other women, and the men who did such things were other men.

None of them were the men we knew. The newspaper stories were like dreams to us, bad dreams dreamt by others. (66)

Once instituted, silence and passivity are no longer optional; they become mandatory and are inflicted on the subservient, the fearful, the naïve by undeserving opportunists who easily climb the social and political ladder in turbulent times. If the gender element is added to the mix, the Gileadean community of *The Handmaid's Tale* – with men in command and women enslaved – is the immediate outcome. A way out, paradoxically, is through dreams which summon past realities (pleasant this time), as well as meditations on what had previously gone unseen or had been misunderstood. The future is not even imagined. The journey backwards happens at night, in the sections specially dedicated to the protagonist's silent retrospection (seven of which are entitled *Night*, thus being ironical towards the patriarchal stereotyping of femininity).

Otherwise, dreaming is forbidden. Reality needs experiencing; learning and practising; forgetting. Therefore, the rest of the fifteen sections form a collage of distopic images, zooming in on everyday daytime activities enforced on women, in sharp contrast to the "evil" ways of the old world:

- *Shopping* – introduces counter stereotype shopping rituals, with women punished to only buy what they are told, from places like "Lilies of the Field" – selling habits, "a good word" for dresses, since "habits are hard to break" (34) –, "Milk and Honey" or "All Flesh" – selling basic produce, dairy and meats (15–44);
- *Waiting Room* – narrows down on Offred's assigned accommodation, at one of the commanders, where she is expected to wait to be fed, to be told what to do; in her words: "I'm waiting in my room, which right now is a waiting room." (60); "I wait, washed, brushed, fed, like a prize pig." (79)
- *Nap* – adds the torture of compulsory standstill, inducing the desire to do things; anything; this "blank time," "unfilled time," "long parentheses of nothing," "time as white sound" (79–80) is a form of dying, efficiently used as a tool in the re-education of women;
- *Household* – shifts the focus to the house they are all living in: "Household: that is what we are. The commander is the head of the household. The house is what he holds. To have and to hold, till death do us part." (91); it includes details on the staged rape "ceremony," so horrendously graphic that sarcasm is used to soften it down: "I remember Queen Victoria's advice to her daughter. *Close your eyes and think of England*." (105)
- *Birth Day* – centres on another ceremony, an all-female one, of giving birth in public, amid orchestrated chants from "supporting" wives and

handmaids: "'Breathe, breathe,' we chant, as we have been taught. 'Hold, hold. Expel, expel, expel.' Five in, hold for five, out for five.'" (133)
- *Soul Scrolls* – opens the discussion on technology and fundamentalism; *Soul Scrolls* is a store with shatterproof windows, displaying "print-out machines […] known as Holy Rollers." (175) "What the machines print is prayers. […] There are five different prayers: for health, wealth, a death, a birth, a sin." (176)
- *Jezebel's* – enters the world of 'fallen women', the equivalent of a brothel, a work camp slightly less punitive than the Colonies (areas of radioactive waste): "They said I would be a corrupting influence. I had my choice, they said, this or the Colonies." (261)
- *Salvaging* – tells the story that does not want to be told: of the ceremony where women participate in the hanging of other women, punished for "lesser" crimes (unchastity, adultery, attempted escape, the killing of a handmaid): "I've leaned forward to touch the rope in front of me, in time with the others, both hands on it, the rope hairy, sticky with tar in the hot sun." (288)

Everything happens inside the Wall – a physical border of urban Gilead and a panoply of hanged (still hanging) unruly citizens – meant as a reminder of the fact that all hope has had to be abandoned. The dilapidated citadel – whose realistic, naturalistic portrayal seems to be inspired from post-mortem photography – reeks of the putrefaction of Death, although the die has been cast and, "romantically," Life in Death has won the game despite the numerous Prayvaganzas organised.

> The air got too full, once, of chemicals, rays, radiation, the water swarmed with toxic molecules, all of that takes years to clean up, and meanwhile they creep into your body, camp out in your fatty cells. Who knows, your very flesh may be polluted, dirty as an oily beach, sure death to shore birds and unborn babies. Maybe a vulture would die of eating you. Maybe you light up in the dark, like an old-fashioned watch. Death-watch. That's a kind of beetle, it buries carrion. (122)

Beyond it, in the rural areas, lie the Colonies, one of the last circles of this restaged and re-enacted Hell (followed only by concentration/extermination-camp-like furnaces and human experiment sites). The Colonies are set up and used to impose canons, to threaten and terrorize, being central to the indoctrination process carried out in the so-called educational institutions (i.e. mandatory subject in the syllabus of the Red Centre, which trains the Handmaids).

> [T]hey showed me a movie. Know what it was about? It was about life in the Colonies. In the Colonies, they spend their time cleaning up. […] Sometimes it's just bodies after a battle. The ones in city ghettoes are worst, they're left around longer, they get rottener. This bunch doesn't like dead bodies lying around, they're afraid of a plague or

something. So the women in the Colonies there do the burning. The other Colonies are worse, though, the toxic dumps and the radiation spills. […] They say there's other Colonies, not so bad, where they do agriculture: cotton and tomatoes and all that. But those weren't the ones they showed me the movie about. (260)

The dense political, historical and social implications may be read in the references to red totalitarianism and far-right dictatorship; to slavery, warfare and the Holocaust; to exploitation and death; to poverty, famine and epidemics; to hazardous waste and ecological disaster. The gender bias indicator is nonetheless the most emphatic. It is the women-cleaners who are sacrificed; they sweep and mop the mess; they are made responsible. They, too, have to be submitted into 'bearing fruit' to keep the mechanism going and be thankful for having been offered the privilege. The rebel or ungrateful ones are dehumanized; investigated in laboratory conditions, they are then disposed of: "They won't even bother to send you to the Colonies. You go too far away and they just take you up to the Chemistry Lab and shoot you. Then they burn you up with the garbage like an Unwoman." (228)

All these concentric settings, allegorically represent the metamorphosis of June into Offred, the journey from a Woman to an Unwoman. They are reluctantly woven into the protagonist's narrative, reinforcing the messages of the plot and of its numerous subplots, while facilitating the close exploration of the frontiers of actuality and rationality. The nightmare she is living is juxtaposed to the dreams of who she once was, and a return to reality is conjured. The central character assumes the task of storyteller as a means of survival (while committing the crime of hoping), of remaining sane, and not as an effort to document a segment of her life. Her belief is in the supreme power of the word, in the freedom of speech – theoretical notions, however; arbitrary and hypothetical respectively. Slipping out of control. "I would like to believe this is a story I'm telling. … If it's a story, then I have control over the ending. Then there will be an ending, to the story, and real life will come after it. […] It isn't a story I'm telling." (49)

Doing the only thing she can to intervene in the course of events – reporting on her life in Gilead – Offred is handling a double-edged sword. What she documents, though real, is destined to be read as fiction. Since the inconceivable is already taking place, its rendition cannot but be interrogated based on fact. As for the act of communication, it is rendered ineffective by Offred's coercion into silence and by her imaginary addressee. "By telling you anything at all I'm at least believing in you, I believe you're there, I believe you into being. Because I'm telling you this story I will your existence. I tell, therefore you are." (279)

Her words may very well be read as a covert metafictional statement that Atwood aptly introduces, apparently to stir reactions from the public – generally uninterested in contemplating the subtle criticism of dystopian writing, therefore unwilling to implement change or too idle to stop the ongoing destruction of what humanity has inherited and managed to achieve.

Paratextual ending: "Historical Notes"

Ironically, Offred's audience takes material shape in the novel's overt metafictional aside, "Historical Notes". Although it essentially invites at re-reading the novel and probes into how the present announces the future and how history is silent on her story, "Historical Notes" practically misreads/unreads everything, and proves somehow incapable of 'translating' the open ending which is projected against its very own reality, some two hundred years post the apocalypse reported: "And so I step up, into the darkness within, or else the light." (307)

This paratext (and its embedded paratext)[25] is advertised as a "partial transcript of the proceedings of the Twelfth Symposium on Gileadean Studies held as part of the International Historical Association Convention, which took place at the University of Denay, Nunavit [*deny none of it*], on June 25, 2195" (311) which authenticates the reality in/of fiction as captured by the main narrative. Significantly, it reinforces the data on the rate of infertility in Gilead, scientifically accounted for in this conference which seeks to shed light on the environmental conditions that allowed the instauration of this monstrous form of government.

> Some of the failure to reproduce can undoubtedly be traced to the widespread availability of birth control of various kinds, including abortion, in the immediate pre-Gilead period. [...] Stillbirths, miscarriages, and genetic deformities were widespread and on the increase, and this trend has been linked to the various nuclear-plant accidents, shutdowns, and incidents of sabotage that characterized the period, as well as to leakages from chemical and biological-warfare stockpiles and toxic-waste disposal sites, [...] and to the uncontrolled use of chemical insecticides, herbicides, and other sprays. (316–317)

Apart from that, "Historical Notes" questions the genuineness of the overall narrative discourse, mostly due to the fact that it was assembled by a female author/narrator. Offred's account of the tyrannical Gilead is also derided, dismissed as subjective and uncreditable on the basis of its orality. Her spoken

---

25 More details on the texture and architecture of the ending paratext may be found in Michaela Praisler and Oana Gheorghiu, "The Art and Politics of Rewriting. Margaret Atwood's Historical Notes on *The Handmaid's Tale*" (2019).

testimony, recorded on tape following her escape from the Waterford household, is additionally accused of being contaminated by the memory of the events narrated, indirectly told from a spatial and temporal distance. In short, the academia, portrayed in the "Historical Notes" as a patriarchal zone, almost as a mini-version of Gilead, sends the urgent matter of environmental crises and the condition of women to the background, laying emphasis on the power of its truth-constructing apparatus (Foucault 1972), and on inscribing the past with new meanings through its menacing scientific research, disseminated at conferences and in published papers. In the writer's words, "that's what happens to ustopian societies when they die: they don't go to Heaven, they become thesis topics." (*The Guardian*)

Conclusion

The final fictional warning on the world and the word, on the erasure of topics related to women and the environment that Atwood formulates in the "Historical Notes on *The Handmaid's Tale*", unfortunately, does not seem to have reached a very large (or very enthusiastic) public. Which somehow explains her decision to engage in open debates on women's rights and in environmental activism via multiple media, thus also addressing numerous others besides the avid traditional literature readers. Hopefully, the academia too will leave their ivory tower to visit the old house society is inhabiting – whose layout reveals upper male exclusivist clubs and lower female servants' quarters, and which is now at the mercy of natural disasters. Ideally, sooner rather than later, they will help rethink the blueprint of the new one (still) being built as we speak.

References

Atwood, Margaret. "The Road to Ustopia". *The Guardian*, 14 October 2011. Web.

Atwood, Margaret. *The Handmaid's Tale*. London: Vintage, 2010. Print.

Channel 4 News. "Margaret Atwood: 'There is a climate crisis'". Interview with *Krishnan Guru-Murthy*. *Facebook*. 11 September 2019. Web.

Cixous, Helene. "Sorties". 1975. *Modern Criticism and Theory*. Second Edition. Ed. D. Lodge and N. Wood, Harlow, UK: Pearson Education, 2000, 264–270. Print.

Foucault, Michel. "Truth and Power". 1972. *From Modernism to Postmodernism. An Anthology*. Ed. L. Cahoone. Oxford: Blackwell, 1996, 379–380. Print.

France 24, Encore! "Margaret Atwood on Women's Rights and the Climate Crisis". Interview with Janira Gomez. *YouTube*. 20 February 2020. Web.

Gaard, Greta (ed.). *Ecofeminism: Women, Animals, Nature*. Philadelphia: Temple University Press, 1993. Print.

Mellor, Mary. "Gender and the Environment". *Ecofeminism and Globalization. Exploring Culture, Context, and Religion*. Ed. H. Eaton and L. A. Lorentzen, Maryland: Rowman & Littlefield Publishers Inc., 2003, 11–22. Print.

Nexus Instituut. "Margaret Atwood – On Fiction, the Future and the Environment. An Informed Dialogue about the World of Ideas". Interview with Eveline van der Ham. *YouTube*. 21 December 2012. Web.

Praisler, Michaela; Oana Gheorghiu. "The Art and Politics of Rewriting. Margaret Atwood's *Historical Notes on The Handmaid's Tale*". *Cultural Intertexts*, vol. 9/ 2019, Ed. M. Praisler et al., Cluj: Casa Cărții de Știință, 2019, 171–181. Print.

Thorpe, J. R. "What Exactly Is Ecofeminism?" *Bustle*, April 22, 2016. Web. 8 September 2021.

Zizek, Slavoj. *Welcome to the Desert of the Real*. London and New York: Verso Books, 2002. Print.

Andreea Cosma

Ecopoetics in Diane di Prima's *Revolutionary Letters*

Abstract: This chapter is concerned with the ecopoetics of space in di Prima's *Revolutionary Letters* (1968) and with the analysis of real, imagined and ideological spaces, described as "metamanifest places," that foster a sense of social and environmental progress combined with aversion towards the oppressive authority and ideology of the time. I argue that di Prima's poems may be read as a literary manifesto against capitalist structures, with a view to positioning women as ideological and environmental activists.

Introduction

The controversial relationship between the natural environment and the built environment has been a debating point since the Industrial Revolution and, later on, during the development of technology, rise of the globalized society and corporate era and, which became increasingly visible after the Second World War. The emergence of fields of study, such as Geocriticism and Environmental Criticism, starting with the 1960s aimed at raising awareness on the importance of the people-place conjunction, but also to the negative impact of pollution and environmental exploitation. While the environmental turn manifested itself mostly during the 1980s, the pro-environment movement was pioneered by writers, artists and activists that witnessed World War Two, who could picture the destruction of both, the natural and the built environment. While the first volume of the *Revolutionary Letters* was published in 1969 by Long Hair Books, the fifth and last edition was published in 2021 by City Lights publisher. With each new edition, di Prima added new letters to the volume in order to portray socio-political, cultural, and environmental aspects about each era. Therefore, her *Revolutionary Letters* serves as her long-term work and also as a documentation of the last fifty years. This chapter is concerned with geo- and ecocritical readings of di Prima's first forty letters, which correspond to the post-war period. The messages of revolt and activism behind di Prima's work lie in the analysis of places which the poems reveal, that could be called, I argue, "metamanifest places[26]."

---

26 The term "metamanifest place" refers to the joint meanings of real, geographical space, its metaphoric depiction and its ideological representation. These constructions place

In Lawrence Buell's *The Future of Environmental Criticism* (2005), it is noted that although "ecocriticism" serves as an umbrella-term for studies that tackle literary and environmental works, it is "environmental criticism" that better defines the spectrum through which, both natural and constructed environments can be approached in literary works (viii). Buell further notes that the "literature-and-environment studies" has always tackled the real sciences and the humanities together and that it is in recent years that this field has been approached by cultural studies scholars. The relationship between culture and nature has always been drawn by the human presence. It is creativity and politics that turn nature into a cultural phenomenon and one may distinguish between environmental manifestos and nature-inspired art. While they fully complement each other toward the achievement of nature-conservation and nature-appreciation, they are created from two different perspectives: a political one and an aesthetic one.

In his *The Song of the Earth* (2000), Jonathan Bate discusses these distinctive approaches in environmental literature and compares ecopoetics with ecopolitics. Bate states that ecopolitics refers to direct instructions on how to dwell with the earth but ecopoetics is about experiencing it: "Reverie, solitude, walking: to turn these experiences into language is to be an ecopoet" (42). In another one of Lawrence Buell's works on environmentalism, *Writing for an Endangered World: Literature, Culture, and Environment in the U.S. and Beyond* (2001), it is stated that there are four types of reader-engagement as regards the literature for and about the environment:

> They may connect readers vicariously with others' experience, suffering, pain: that of nonhumans as well as humans. They may reconnect readers with places they have been to and send them where they would otherwise never physically go. They may direct thought toward alternative futures. And they may affect one's caring for the physical world: make it feel more or less precious or endangered or disposable. (2)

While readers may grow fonder of a certain literary work based on one or more of Buell's reasons mentioned above, writers may intentionally design the spaces or cartographies of their texts in order to be read as "metamanifest places," which transform the literary work into a tool that generates awareness and may create an impact on the world. As Robert Tally mentions in his *Spatiality* (2012), "with respect to literary and cultural productions, these spaces call for new cartographic approaches, new forms of representation, and new ways of imagining

---

the author in the shoes of the activist and emphasize the cartography of the literary work as a space of protest.

our place in the universe" (42). Space is a mosaic that is in a continuous process of transformation and it is made up of an infinite number of pieces, each one representing a mental map. Soft cities (mental maps) and hard cities (official maps) (Dear 9) no longer necessarily need to be analyzed on layers, but rather by putting the pieces together, as the real and imagined space combined offer a full and thorough understanding of a place. Metamanifest places are concerned with putting together pieces of what Tally describes as literary cartographies (the writer's maps) and literary geographies (the reader's maps) (Tally 3), with a view to revealing the activist dimension of a text concerned with environmental politics.

Diane di Prima tackles the topic of the environment from both an aesthetic and a political view, supporting the view according to which the poet's volume illustrates her work as both artist and activist. Di Prima's ecopoetics thoroughly convey the space that surrounds her, as well as people and places that were meaningful to her and that led to her preoccupation with environmental protection. The poetic voice from the *Revolutionary Letters* seems to be fully engaged in the described experiences, which turn di Prima's poems not only into political statements, but also into authentic artifacts devoted to the natural environment.

Post-WW II, women and the tribe

Di Prima's volume begins with "April Fool Birthday Poem for Grandpa," which sets the tone for the following ninety-three "Revolutionary Letters." While part of the volume is concerned with "know-how" information that instructs the reader on how to react to, and act during, a revolution, the rest of the volume seems to describe an imagined space in which people protect and celebrate the environment. The first poem of the volume, which represents a tribute to her grandfather, alludes to the early influences that shaped di Prima as a poet:

> thank you
> for honestly weeping in time to
> innumerable heartbreaking
> Italian operas for
> pulling my hair when I
> pulled the leaves off the trees so I'd
> know how it feels (5)

The poetic voice alludes to her grandfather's teachings about how to appreciate both art and nature. The sensibility of the poetic voice's experience in each childhood lesson expresses a drive to nature and, I argue, turn di Prima into an ecopoet.

She does not only write about the need to preserve nature, but also about heritage and the community, creating a revolution of love rather than a green revolution. To di Prima, it is love of people, places, tradition, culture and nature that leads to peace and abundance. She also emphasizes the importance of communal values and imagines how proud her grandfather would be of her artistic and political agenda and circle:

> young men with light in their faces
> at my table, talking love, talking revolution
> which is love, spelled backwards, how
> you would love us all, would thunder
> your anarchist wisdom at us, would thunder
> Dante, and Giordano Bruno, orderly men (6)

The poetic voice feels as if she were taking further the grandfather's legacy while, at the same time, she tries to preach the teachings of the "orderly men," like Dante, who revolutionized poetry, and Giordano Bruno, who challenged the way people thought about the universe. It is love and empathy for one's surroundings and community, it is also going against the norm that di Prima's revolution is about. The poetic voice feels that she and the "young men with light in their faces" are part of a greater quest, which started hundreds of years before.

"Revolutionary Letter #1" starts the series of letters in the volume, by emphasizing the power of individual action: "I have just realized that the stakes are myself/ I have no other/ ransom money, nothing to break or barter but my life" (8). It is a statement of freedom and, at the same time, one that empowers the reader to make their own decisions and to be responsible for them. "Revolutionary Letter #2" continues as an ode to the power of the community: "we are/ endless as the sea, not separate, we die/ a million times a day, we are born/ a million times, each breath life and death" (9). The "tribe" that di Prima mentions symbolizes human life as a global construct, alluding to the connection that we all have through life and the environment.

"Revolutionary Letter #3" could be read as a guide on how to adapt to a crisis. While some of the letters describe a utopian-like place where people manage to co-inhabit in peace and understanding, the third letter describes a dystopian-like situation, where people have to quarantine, store food for a long period of time and adapt to live without basic necessities: "make a habit/ of keeping the tub clean and full when not in use/ change this once a day, it should be good enough/ for washing, flushing toilets when necessary" (10). Di Prima's poem seems to be inspired by the 1967 Newark riots: "at the first news of trouble: they turned off the water/ In the 4th ward for a whole day" (10). Throughout the poem, the poetic voice describes the atmosphere of anxiety during the riot (i.e. revolution) which is described as war-like due to the straight-forward instructions on how to prepare for, and survive during, difficult times.

Diane di Prima shifts between the dystopian and the utopian throughout her letters, which could be interpreted as a way to express hope for a better future during the Cold War and Civil Rights tensions of the time when she was writing the poems. In "Revolutionary Letter #4," the ecopoetics of space that di Prima uses conveys a picture of unity with nature:

> We return with the sea, the tides
> we return as often as leaves, as numerous
> as grass, gentle, insistent, we remember
> the way,
> our babes toddle barefoot thru the cities of the universe. (14)

The images of nature as a nurturing mother and the "cities of the universe" as the space of life cycles create a "metamanifest place" which promotes peace and a non-violent lifestyle.

The 1960s mark a period when ecology and feminism start marching together. Women struggle to escape social and economic pressures while ecologists try to educate the mainstream society about the consequences of developing machinery, products and industries that are not environmentally-friendly. As Carolyn Merchant notes in her book, *The Death of Nature: Women, Ecology and the Scientific Revolution* (1980), the joining of the two movements, the Second Wave Feminism and the Environmental Movement, led to the creation of a new understanding of society, which is not centered on patriarchal principles, such as gender roles and economic expansion, but rather on a social and economic agenda in which gender, race and environmental issues were brought up. The aim was, as the poetic voice declares, to obtain the "full expression of both male and female talent" and "the maintenance of environmental integrity" (xix).

Moreover, di Prima dedicates "Letter #44" to her "sisters" noting that "we/ liberate, and nourish, as the earth" (106). The ecopoetics in this letter creates a sense of resemblance between women and the earth, portraying them as the nurturing part of society that give stability and a sense of belonging to the community, who are oppressed despite their positive contribution. Women hear "the plea in the voices around us, not words/ of passion or cunning, discount, anger or pride" (106). Another voice of the decade, though not a poet, Betty Friedan, discusses in her celebrated study "*The Feminine Mystique*" (1963) the need for women to find their own voice.

> They [women] had to prove that woman was not a passive, empty mirror, not a frilly, useless decoration, not a mindless animal, not a thing to be disposed of by others, incapable of a voice in her own existence. (Friedan 104).

The recurrent question that Friedman raises for her female readers is "who are you?," referring to what one's identity is, besides the one assigned by patriarchal norms. Di Prima answers this question in her letter, saying that women are the "liberators" and referring to the fact that it is women's empowerment that can win the revolution and put an end to social and environmental degradation.

While "Revolutionary Letter #3" prepares the reader with information about logistics during a crisis, such as hoarding food, managing water, obtaining fire and heat sources, "Revolutionary Letter #5" tackles the psychological dimension of a dystopian situation: "you may be called upon/ to keep going for several days without sleep;/ keep some ups around, to be/ clearheaded, avoid 'comedown' as much as possible" (16). The instructions mix the use of organic and chemical products that would help the reader cope with stress and lack of sleep. "Revolutionary Letter #6" tackles the idea of racial and gender-based oppression:

> they love us and want us to practice birth control
> they love us and want the Hindus to kill their cows
> they love us and have a colorless tasteless powder
> which is the perfect synthetic food... (19)

The revolution is not only at the environmental level, but also at the cultural and social ones. It is the freedom of choice for women to decide for their own bodies and for people to follow their faith. The space that di Prima describes is also one in which food is processed and creativity and the joy of life disappear.

The imagined space in "Revolutionary Letter #7" is compared to a structure in which everyone is "tunneling under:" "friend and foe, like a million earthworms/ tunnelling under this structure" (21). Just as the earthworms make their way around the structure, the citizens try to make ends meet around the city. It is di Prima's way of expressing the fact that humans are part of a whole natural universe and that there is a connection and a pattern between humans and the nature around them. The cartography of the revolution starts to take shape in "Revolutionary Letter #8," as the poetic voice instructs the reader to choose the terrain of the battle ground in their favor (23): "don't let them lure you/ to Central Park everytime" (23) and then tests the reader on their strategy to survive in New York: "Central Park West, or Fifth Avenue, which would you/ choose?" (23).

Di Prima continues with the political agenda behind the battle in "Revolutionary Letter #9," which seems to be based on an ideal, even mythicized notion of "common wealth," a "tribe" based on socialist values: "no one 'owns' the land/

it can be held/ for use, no man holding more" (26). Di Prima fantasizes about an American society with no debts and no consumerism and no social Darwinism, but a sharing community: "and what you make/ above your needs be given to the tribe" (26). The metamanifest place that di Prima contours in this letter is a space of primitivism and ideal socialism, perceived even by the poetic voice as a fantasy or as a system that is impossible to work in reality. However, the reader is encouraged to take a different approach on life and politics: "None of us knows the answers, think about/ these things" (27). As studies show, tribal settings differ significantly throughout the world: while it may seem that tribes manage to live according to communal values and a genuine socialist mindset, it has been proven that power-structures and ownership are pressing issues among the inhabitants (Sprinzak 629).

The space in "Revolutionary Letter #11," moves from New York to San Francisco, where the poetic voice meets the "Diggers," (or "do-gooders," as a 1967 *Time* article calls them), a left-wing-group of local theater artists who were offering shelter and food to the needy, usually, minorities, hippies and Beats. The poem focuses on the way the hippies are stigmatized at a national level due to the media: "SMASH THE MEDIA, I said,/ AND BURN THE SCHOOLS" (31). Di Prima addresses the root-causes of social conformity and oppression, the school as the shaper of the young generation and the media, as the manipulator of the masses. The author raises the same problem in "Letter #19," by emphasizing that the media, "whose radiant energy/ kills brain cells, whose subliminal ads/ brainwash your children, have taken over/ your dreams (47) and that the school is where "all our kids are pushed into one shape" (46).

Space, race and the environment

The metamanifest place of "Letter #11" expresses an image of decay, "across fields of insecticide and migrant workers" (30), where uncontrollable forces challenge di Prima's revolution, people struggle to survive poverty while the environment is being poisoned. "Revolutionary Letter #12" discusses the consequences of social change: "the vortex of artistic creation is the vortex of self-destruction/ the vortex of political creation is the vortex of flesh destruction" (32). The revolution requires a space of creation, which may impact the artist through destructive feelings and habits due to the overwhelming and cathartic nature of the creative process. At the same time, political and social progress polarizes the society, creating space for struggle and conflict. Letter "#13" emphasizes the irreversibility of man-made destruction:

> just gather spirit, see the forest growing
> put back the big trees
> put back the buffalo
> the grasslands of the midwest with their herds
> of elk and deer
> put fish in clean Great Lakes
> desire that all surface water on the planet
> be clean again. (34)

The ecopoetic dimension of this letter creates a sense of dislocation, the Earth is no longer what it used to be, because man has taken over the environment. The "forest," the "grasslands of the Midwest," and the "Great Lakes" create a cartography of metamanifest places of the United States' natural environment as a tribute to extinct species. The feeling of guilt that the poetic voice expresses has also been observed by Estíbaliz Encarnación-Pinedo, in her work, "On Webbed Monsters, Revolutionary Activists and Plutonium Glow: Eco-Crisis in Diane di Prima and Anne Waldman" (2021). According to her, "Letter #16" is a case in point, because it opens up with a direct *mea culpa*" (Encarnación-Pinedo 3). The line "we are eating up the planet" (di Prima 40), seems to position the self in line with all people and things that affect the Earth. Later in the poem, however, it is clarified that "we" stands for the United States: "5% of the world's people [i.e., the United States] uses over 50% of the world's goods" (di Prima 41). The poet further notes how the country is exploiting the environment globally: "the New York Times/ takes a forest, every Sunday, Los Angeles/ draws its water from the Sacramento Valley/ the rivers of British Columbia are ours/ on lease for 99 years" (40). The cartography of polluters creates a metamanifest place which points out at corporations and metropolises for being environmentally irresponsible. Economic expansion and globalization also play an important role in the way in which infrastructure and consumerism overwhelm the environment. As di Prima writes in "Letter #17," "The planet will not bear it" (42).

In "Letter #21," ecopoetic concerns are represented by the question of ownership and consumerism: "will the air belong, as it gets rarer?/ the american indians say that a man/ can own no more than he can carry away/ on his horse" (50). The poetic voice encourages a humble attitude towards the environment, as a means of overcoming human greed. "Letter #22" continues with the idea of people's connection to the environment, while considering themselves a part of it. The poem opens with the question: "what do you want/ your kids to learn?" (51). The ecopoetics used in the following questions, raised in the poem and addressed to the reader, aim to show the shallowness of the mainstream society, who forgets that, like all other beings, its heritage is rooted in nature and not in concrete or plastic: "what is he doing all his learning years/ inside, as if the planet were no more than a vehicle/ for carrying our plastic constructs around the sun" (52). The poetic voice calls the reader to return to nature, learn and embrace the environment.

In "Letter #32," the poetic voice emphasizes the relation between race, privilege and power: "not killing a few white men will bring/ back power, not killing all the white men, but killing/ the white man in each of us" (81). It is not only the color of one's skin that affects one's actions, but the values in which one is raised. "Killing" the white man inside us refers to giving up on commodities, pleasure, privilege and entitlement. The ecopoetic interpretation of di Prima's letter reveals that "white man" is synonymous with "greed," because, the voice says, "white man" sits on the highest levels of the social ladder. Further in the poem, the voice refers to the city, white men's creation, and points out their disruptive character on the environment: "show me/ a city which does not consume the air and water for miles around it" (82). Through the ecopoetics used, it is pointed out that one should focus on the "flesh" and the "spirit of man" rather than creating a city, as it is the individual and the community that matter most. In "Letter 33" the poetic voice raises the question: "how far/ (forward is back) are we willing to go/ after all?" (84). She considers that future means degradation at a spatial and environmental level, which affects heritage and creativity. The "Cybernetic civilization" may or may not "save the water," however, it will not teach people about their roots.

Di Prima's revolution aims to obtain both, the preservation of nature and of humanity. She believes that they can be reached through the creation of villages rather than cities. Her utopian village is mapped in the letters and it consists of "tepees," "domes" and "hogans," the cars are turned into "flower pots or sculptures," the metal is put back into the earth and the petroleum lines "BLOW UP" ("Revolutionary Letter 34" 86). While "Letter #34" invites people to join the revolution: "hey man let's make a revolution" (85), "Letter #35" serves as an urge to "rise up," because "the earth cries out for aid" (87). The poetic voice takes the position of a leader in "Letter #35," stating that "we must reclaim/ the planet, re-occupy/ the ground" (88). The recurrent use of the pronoun "we" gives the impression that there is a community of people that the poetic voice addresses. While in "Letter #16" the poetic voice feels as if it were part of the ones to blame for the degradation of the Earth, "Letter #36" dissociates "we" from "they:" "who is the we, who is/ the they in this thing" (89). After contemplating on this binary opposition, "did we or they kill the Indians?" (89), the reader is urged to revolt and to reject "a share of the guilt" (89). David Harvey claims in his "*Rebel Cities*" (2012) that "social movements often fall prey to severe factionalism and infighting" (146). However, di Prima's revolution seems to base itself on grassroots efforts and guerilla manifestations, fighting against precise target groups while, at the same time, it is meant as a way of life rather than an actual confrontation with the enemies.

"Letter #37" reveals di Prima's literary cartography of the USA, mapping the country from the "east edge" which is "megalopolis" to the "west edge" where "the sisters/ raise their bastard young on welfare checks & rotten/ sprayed vegetables" (91). The image of decay moves closer to the city in "Letter #39," when New York is mapped by "Tompkins Square Park" and the "Natural History Museum" with small patches of nature still existing in the city. "Letter #40" illustrates how "the sea licks its chops/ at the oily edges of Los Angeles (98)". Both maps, of New York and Los Angeles, depict metamanifest places that stand behind di Prima's revolution. It is the poisoning of waters, the caging of the birds and the oppression over nature that fuel the poet's cry for a change. The anti-war movement is also present in di Prima's Works, as "Letter #54" states:

> HOW TO BECOME A WALKING ALCHEMICAL EXPERIMENT
> eat mercury (in wheat & fish)
> breathe sulphur fumes (everywhere)
> take plenty of (macrobiotic) salt
> & cook the mixture in the heat
> of an atomic explosion (126)

"The experiment" that the poetic voice describes, refers to the poor quality of mass-produced foods and the consequences of buying such products that large industries make profit of. "The experiment" culminates with the use of the "atomic explosion" in order to emphasize that people are only the casualties of economy-driven agendas.

"Letter #60" analyzes the map of large cities in the United States and points out the effects of the "urban renewal" (138). The gentrification of the suburbs leaves the poor and the discriminated: "out of sight, out of mind" (138). Some of such neighborhoods from San Francisco, New York and Boston are mentioned: "Hunters Point, Lower East Side, Columbia Point," calling them "remote & indefensible piece of ground" (138). These metamanifest places in the poem, such as "the slums from the heart of town" (138) portrays the American city as one that is only money-oriented rather than based on values of organic development and social growth. The poetic voice seems hopeless and disappointed at people being passive about these changes by saying that "few will see, & fewer will object" (138). City-exploitation is also tackled in "Letter #61:" "off-shore drilling renewed, Santa Barbara & elsewhere we can expect" (139). Di Prima sees consumption as a snowball: industries around the city start exploiting the surrounding lands to make a profit, the more industries a city has, the more gentrified it becomes, causing the city to become more a money-making machine and less a community.

Conclusion

The first poem, this volume with, emphasizes di Prima's reasons behind the revolution. The first twenty letters shift between utopian and dystopian spaces, alternating between places in which the poetic voice feels paranoid and anxious and then hopeful and dreamy. The dystopian places portray a situation in which people need to hoard supplies, learn how to fight and defend themselves or lock themselves inside, while there is destruction and death outside. The utopian places depict communities of people, sharing everything, being connected to nature and living humble and happy lives, while the Earth is thriving. The first ten letters shift from warnings and instructions to promises and fantasizing. Starting with letter eleven, the poetic voice begins to call the blame on various factors, destroyers and polluters, seeming as if the voice is demanding that somebody takes responsibility for their actions. In the thirteenth letter, the poetic voice becomes aware of the irreversibility of the situation and accepts that even she is part of the factors that affect the environment by using the pronoun "we." From the sixteenth to the twenty fourth letter, the poetic voice changes its strategy in alluring the reader to become protective of the environment through persuasion, by trying to make him/her empathize with, and relate to, the situation. From letter twenty fifth onwards, the poetic voice raises the question of leadership, strategy and battle, emphasizing that it is for the future generations that we need to fight pollution, discrimination and racism.

Di Prima's *Revolutionary Letters* is an ongoing discourse on the intersections of gender, race, space and environment. Her messages tackle injustice, analyze root causes of the existing problems at the time and identify solutions. While the letters approach the reader through different strategies in order to convince them to join the revolution, the poet remains confident that the texts will reach the people, "the word has power, the chant is going up" (99), as she knows that the revolution is a natural "turn" in life, "Revolution: a turning, as the earth/ turns, among planets" (100). Di Prima's revolution involves the empowerment of women, the power of the word and the fact that it is a revolution in one's lifestyle rather than expressed through a battle. The analysis of ecopoetics and metamanifest places in di Prima's *Revolutionary Letters* reveals the end-aim of her fight, that is the reconnection between humans and the environment, which will naturally solve all the other social problems of the current times, such as poverty, greed, discrimination and pollution. As portrayed through the letters, people can only relate to and resonate with the environment through genuine empathy and devotion for their surroundings. Di Prima's letters, which represent her relentless drive for creation and growth throughout her life as a poet and as an activist, serves both as a "how-to" guide for surviving the "cybernetic civilization" and as a book of teachings about how to appreciate the environment with all that it comprises and how to contribute to the "tribe's" existence, that is the community one lives in.

References

Bate, Jonathan. *The Song of the Earth*. London: Picador, 2001. Print.

Buell, Lawrence. *The Future of Environmental Criticism: Environmental Crisis and Literary Imagination*. Oxford: Blackwell, 2005. Print.

———. *Writing for an Endangered World: Literature, Culture, and Environment in the U.S. and Beyond*. Cambridge: The Belknap Press of Harvard University Press, 2001. Print.

Dear, Michael J. "Creativity and Place." *GeoHumanities: Art, History, Text at the Edge of Place*, ed. Michael Dear. New York: Routledge, 2011. 9–18. Print.

Di Prima, Diane. *Revolutionary Letters*. San Francisco: Last Gasp, 2007. Print.

Encarnación-Pinedo, Estíbaliz. "On Webbed Monsters, Revolutionary Activists and Plutonium Glow: Eco-Crisis in Diane di Prima and Anne Waldman." *Humanities*, vol. 10, no. 4, 2021. 15–28. Web.

Friedan, Betty. *The Feminine Mystique*. New York: W. W. Norton & Company, Inc, 2001. Print.

Harvey, David. *Rebel Cities: From the Right to the City to the Urban Revolution*. New York: Verso, 2012. Print.

Merchant, Carolyn. *The Death of Nature: Women, Ecology, and the Scientific Revolution*. New York: Harper & Row, 1989. Print.

Sprinzak, Ehud. "African Traditional Socialism – A Semantic Analysis of Political Ideology." *The Journal of Modern African Studies*, vol. 11, no. 4, 1973. 629–647. Web.

Tally, Robert T. *Spatiality*. London: Routledge, 2013. Print.

"Youth: The Hippies." *Time*, vol. 90, no. 1, July 7, 1967. 1–10. Web.

Ludmila Martanovschi

Ecodramaturgy Meets the Arctic: Chantal Bilodeau's *Sila* and *Forward*

Abstract: This study demonstrates that Chantal Bilodeau's *Sila* (2015) and *Forward* (2017), the first two plays in a series that exposes climate change in the countries of the Arctic Circle, focus on essential themes for ecodramaturgy. In these two works, the writer-activist explores gendered responses to the environmental crisis over several decades, reflecting on the artist's role in helping audiences visualize it.

Introduction

Award-winning playwright based in New York City, Chantal Bilodeau, is a vibrant voice in environmental activism, having initiated artistic projects with far-reaching reverberations globally. One of them is the Arctic Cycle, a series of eight plays set to examine the effects of climate change in the eight countries of the Arctic[27]. Focusing on *Sila* (2015) and *Forward* (2017), the first two plays of the cycle, this study demonstrates that the works under consideration illuminate the most relevant themes and techniques of ecodramaturgy and raise awareness about current ecological issues in impactful ways, even if their subject matter is distinct: *Sila* (Canada)[28] features an Inuit mother who loses her adolescent son in the context of contemporary pressures placed on their indigenous community, the human story paralleling that of a female polar bear and her cub, while *Forward* (Norway) looks at icescapes, an Arctic explorer and his complicated legacy.

Given the fact that the playwright is programmatically exploring a new cultural milieu with each play, critical attention is paid here to whether she manages to voice local worldviews and concerns without risking (mis)appropriation. Furthermore, the current analysis reflects on the ways in which the two plays engage with intersectionality, considering the implications of gender, race and class in connection to the climate crisis. Both *Sila* and *Forward* illustrate Bilodeau's approach to what Una Chaudhuri, leading theoretician of space and ecology in the

---

27 The playwright's webpage provides further useful information about this project, referring to the motivation, process and objectives of the series ("Arctic Cycle").
28 Canada is the country the playwright is originally from, which may be one of the reasons why it is featured in the play opening the series.

context of theatre studies, formulates as one of the essential questions that arise nowadays: "How to resist despair while still acknowledging the suffering and losses that undoubtedly lie ahead, for our species as for many others?" (Chaudhuri xvi). Undoubtedly, Bilodeau is one of the artists who seek theatrical solutions to suggest "that it is neither hope nor despair that is most useful now, but rather a new outlook, a new curiosity about how our human selves are related to the more-than-human world" (Chaudhuri xvi). The effective strategies the playwright finds for performances on stage aim at convincing audiences to adopt this outlook that should inform their short and long-term responses to the environmental problems faced by humanity at present.

The advantages theater has in articulating the need for action have often been emphasized. In *Earth Matters on Stage. Ecology and Environment in American Theater* (2021), Theresa J. May insists on the need to give various realities of the climate change artistic visibility and, relying on previous scholarship, establishes: "As a site of culture-making, in which stories are already manifested as action, albeit dramatic action, theater is uniquely positioned to bear witness to this ecological moment, amplify the voices of those impacted by climate change" (239). Indeed, the theatrical performance offers an immersive experience that can transform spectators from indifferent viewers into active participants in debates, projects and practices with positive outcomes. Investigating the objectives and methods employed by theater practitioners and thinkers with a strong interest in ecocriticism[29], Wendy Arons and Theresa J. May provide a definition of ecodramaturgy[30] as "theater and performance making that puts ecological reciprocity and community at the center of its theatrical and thematic intent" (4) in their introduction to *Readings in Performance and Ecology*, a volume they edited as early as 2012. In the same introductory essay, Arons and May also highlight the novelty not only of the "frames for thinking about theater," but also of the "challenges to making theater," (4) including in their volume discussions of recent plays written in the ecodramaturgical vein as well as of eco-conscious approaches to directing classical plays and of efforts for greening theater productions.

The political dimensions of ecodramaturgy become clear with each new play embracing this approach. The representatives of ecodramaturgy, among whom

---

29 Here ecocriticism is used to refer to "the study of the relationship between literature and the physical environment" (Glotfelty xviii), this theoretical approach having gained visibility over several decades.
30 The term itself is actually coined by Theresa J. May in a 2010 *Canadian Theatre Review* article.

Bilodeau is no exception, often convey in visually effective ways matters that have preoccupied environmentalists from all over the world for decades, climate justice being one of them. A paradigm-shifting scholar, Rob Nixon emphasizes the significance of writing as form of protest: "Writing can challenge perceptual habits that downplay the damage slow violence inflicts and bring into imaginative focus apprehensions that elude sensory corroboration" (15). In 2011, Nixon famously defines slow violence as "a violence that occurs gradually and out of sight, a violence of delayed destruction that is dispersed across time and space, an attritional violence that is typically not viewed as violence at all," (2) and details the many challenges this category of violence posits. His book *Slow Violence and the Environmentalism of the Poor*[31] presents the difficulties environmental writer-activists from various disadvantaged areas have faced to date, when attempting to draw public attention to manifestations of violence and injustice. Also, he devotes a full chapter to the contentious points between environmentalism, postcolonialism and American studies (233–262), arguing for the urgent need to "deepen and diversify the dialogue" between these fields in view of "the greening of the humanities" (262). In what follows, Bilodeau's plays are shown to contribute to this desideratum as *Sila* centers on the confluence between environmental and indigenous concerns, while *Forward* probes the Norwegian claims to exploring and exploiting the Arctic.

Sila – The crossroads of environmentalism and indigeneity in Canada

As the printed version of the play testifies, *Sila* was first produced by Underground Railway Theater at Central Square Theater in Cambridge, Massachusetts in April 2014 (*Sila* vii). Recipient of several prizes, it has continued to be staged, attracting attention from both theater goers and critics. This inaugural piece of the Arctic Cycle lives up to the ambitions envisaged by ecodramaturgy practitioners and theorists for performances that should address the environmental crisis in a memorable way, making local concerns visible to a wider audience.

In *Sila*, Bilodeau dares to approach the unapproachable: the death of one's offspring. By relying on an ecofeminist perspective, she interweaves the story of a

---

31   The book continues to be relevant for theater studies scholars who use slow violence in analyzing contemporary plays; one notable example is Marissia Fragkou's chapter "'A Glimpse into Some Other World': Imagining Slow Violence in the Anthropocene" in a book she published in 2019.

human mother with that of a female polar bear, capitalizing on the excruciating sorrow experienced upon losing their children. Contemplating such a reality, clearly presented in the play as a consequence of reckless treatment of the environment, is meant to be the ultimate wake-up call for those still indifferent to the urgency of saving the Arctic. The play emphasizes "interspecies connections" (Arons and May, "Ecodramaturgy" 190), inviting audiences to reconsider the ways in which they think about climate change and its effects. Consequently, the interconnectedness between all breathing elements on earth emerges as a riveting main theme in the play.

Announcing that the setting is "Baffin Island in the territory of Nunavut in the Canadian Arctic" (*Sila* 3) and that several of the characters are Inuit, Bilodeau embarks upon the challenging task of introducing indigenous perspectives without distorting or misrepresenting them. Her mission is to raise awareness about the current struggles of those living in the Arctic area and the need for immediate action, while proving that the geopolitical complexity of the region requires nuanced measures and coordinated efforts. In her "Acknowledgements," Bilodeau thanks the community she worked with, as she puts it – "the people of Baffin Island who welcomed me with warmth, and agreed to share their stories with me," (*Sila* 107) and mentions individuals who answered her questions, thus becoming instrumental in the completion of the play. Among them she names Sheila Watt-Cloutier, an Inuit activist. In her numerous twenty-first century interventions on behalf of the Inuit Circumpolar Commission, Watt-Cloutier defended the Arctic indigenous peoples' rights, explaining that climate change "caused a loss of traditional lifeways, values, and culture, all of which negatively impact the physical and mental health of Inuit communities" (May 252). Many of the actions and principles Watt-Cloutier stands for are taken up by the play, the character based on her being Leanna, a Native woman who can hardly find time for her adult daughter and teenage grandson, given the environmental projects she is involved in.

A key scene in the play captures the moment when Leanna finds out that her petition in which she accused the United States of America of violating human rights was rejected. When her daughter urges her to take into consideration local measures, Leanna replies: "Our hunters can't feed their families, Veronica. Our roads and houses are sinking, and our traditional knowledge is becoming obsolete. No number of educational programs is going to fix that" (*Sila* 27). Leanna's clear-sightedness about evils that cannot be solved by the community alone compels her to act, but the time and efforts invested in fighting for her people, including her family, paradoxically estrange her from her loved ones, daughter

Veronica and grandson Samuel, since she has to travel far and rarely gets to see them.

Leanna's speech before the Inter-American Commission on Human Rights, the actress being instructed to look at the audience (*Sila* 63), opens the second act of the play in a most powerful way. She does begin by talking about rising temperatures and lack of real action to combat high emissions, but then puts aside her notes and declares:

> The real issue has to do with something much more fundamental: our own humanity ... So you may tell me that the world's economic survival is more important than the well-being of a small Arctic nation. You may tell me that anxiety and fear and depression are a matter of personal choice, not of environmental stewardship. You may tell me that drug abuse and ... teenage suicide ...
> *Beat as she fights back tears.*
> ... are by no means a sign of degradation of the Arctic. But I am here to tell you otherwise. (*Sila* 63–64)

Well-informed about the various political positions within the environmental debate in a global context, Leanna capitalizes on the importance of climate justice.

At this point in the play the ravages of contemporary environmental deterioration, i.e. health disorders, physical and mental, with fatal consequences, are not abstractions, but the immediate realities of her own family. Emboldened by the personal meaning her fight has just acquired through the death of her grandson, she continues:

> The real issue is not and will never be climate change. The real issue is that we have lost part of our humanity. We have lost our capacity to care ... The U.S. may or may not recognize a violation of human rights. But unless we open our hearts and embrace not just people we know, but people we don't know, people we will never meet, and people who are not yet even born, we will never value our species enough to make sure it survives. (*Sila* 64)

Her emphasis on the urgent need to adopt a new angle of perceiving the problems that affect the Arctic is meant to convince audiences about the ethical actions that have to be taken sooner rather than later. Through her character's appeal, Bilodeau herself advances a new way of understanding the current environmental crisis: protecting nature is inherently linked to protecting humankind and all other life forms on the planet. More importantly, all humans, irrespective of the geographical area they inhabit or their socioeconomic status, should be granted the right to defend what is essential to their livelihood.

In order to diversify the ways in which the message is put forward, the playwright chooses to have Leanna's daughter, Veronica, express the same concerns about survival and continuance in a new form as she is shown performing spoken-word poetry. She is not only an artist, but a teacher as well. And from the very moment she appears in the fifth scene she charms with her wit, humor and talent. The spoken-word poem *Eskimo Chick*, used in the play with the permission of the author, Taqralik Partridge, is playful and powerful. A love-poem of sorts, it does not refrain from speaking against suicide, one of the cruel realities that indigenous communities face. The addressee, an Inuit girl, is empowered by the first person speaker and urged to celebrate her beauty and strength, embracing life, not death:

> no lie I look you in the eye and see
> a fine line of generations to come
> all sprung from
> your womb – no room
> for contemplating suicide
> suicide, is not the way to go. (*Sila* 24)

Veronica, a fighter for her community through the power of the spoken word, loses the capacity to speak once the news of her own teenage son's death[32] sinks in. Her failed attempts to go on with her art are suggested by her standing in front of a microphone without uttering intelligible sounds. This might be just a dream-like projection, but its purpose is clear in the context of the play. She is shown "desperately trying to form language" (*Sila* 68), this becoming the clearest visual representation of the ways in which grieving for her child can annihilate a mother.

By her side, Tulugaq, a tribal elder[33], connects what he sees with traditional beliefs, explaining that Veronica "seems to have lost some *sila*" (*Sila* 71) and thus diagnosing her condition according to an indigenous worldview. This central concept for the entire play[34] is defined as he speaks in Inuktitut, the translation in English being projected for the audience:

---

32 The play mentions drug abuse, "brought on by the stresses imposed on the native community from loss of identity and sense of place" (Arons and May, "Ecodramaturgy" 190), without clearly stating whether the young man's death was a suicidal act or not.
33 He is a hunter, guide and stone carver; his name means "raven."
34 Megan Sandberg-Zakian, the director of the play's world premiere in 2014, testifies: "we achieved our *sila* through a combination of a lighting effect (a puppeteer slowly reflecting a hand-held halogen light off a pile of colored gel, as if the light itself was breathing) and a recorded soundscape created using the breath of the actors" (iv).

> In ancient tradition, our people believed words were very powerful. Because they were formed with *sila* – the breath, the great life force. When we speak something, that something is given substance. It comes into being … Words are how our individual will take shape. (*Sila* 71)

Relevantly enough, the same concept had been reflected upon earlier in the play in a scene in which the mother bear had taught her daughter the following: "All life is breath. From the original breath that gave us the miracle of Creation to the world itself, *sila* wraps all around us" (*Sila* 43). The indigenous perspective and the animal perspective complete each other since they spring from the same type of relationship to the earth.

While introducing her offspring to basic survival skills, Mama, the polar bear, also empowers Daughter, her cub, to embrace her own destiny, never forgetting that each creature is part of a network: "And with each breath, *sila* reminds us that we are never alone. Each and every one of us is connected to every other living creature" (*Sila* 44). Thus, the lesson of interconnectedness between the land and all the life forms that it nourishes comes from the mother bear. Very importantly, this less conventional character is "played by a puppet or by the same actress who plays Veronica" (*Sila* 3), the play indicating from the very beginning the parallelism between these two mothers. Discussing Bilodeau's representation of "non-human others" in detail, May attests that "the play repositions the polar bears as knowledgeable, feeling beings" (254). Moreover, the polar bears in the play are given "knowledge specific to their lived experience that is interconnected with that of the indigenous people with whom they share place and sustenance" (256). Indeed, this is clearly proven not only by the way both Tulugaq and the mother bear define *sila*, as already explained, but by the way they relate to the dangers posed by the ecological disaster affecting their homeland and livelihood.

When the ice breaks, Mama senses the danger and convinces her daughter to swim to safety by her side. But the distance is quite challenging and soon the former has to carry the cub on her back. Despite all efforts, the young bear's "breath starts to fade" (*Sila* 60), as the stage directions indicate, and she is pulled down to the bottom of the ocean. The audience witnesses the mother's despair and helplessness (*Sila* 61), being prompted to think of all the mothers, non-human and human, suffering such loss as the Arctic ice continues to melt and the environment continues to deteriorate. In the context of the play, Mama and Veronica share the same fate, or, as May suggests, "grieve a world that has cracked open" since "both deaths are caused by climate change" (266).

In his turn, Tulugaq takes a lot of precautions in his dealings with the forces of nature, telling the climate scientist, Jean, who hires him as a guide, that he follows the ways he inherited from his ancestors: "Inuit traditional knowledge. Old learning about living in peace with people, animals, nature. Arctic is not just numbers. Arctic is stories" (*Sila* 52). Here the Native subject is given the agency to define his worldview in his own terms, Bilodeau striving to resist "the Ecological Indian" stereotype, which according to Greg Garrard is an idealization derived or imposed by Euro-American culture from the outside (133). Jean's scientific focus on measurements, capitalizing on rationality and categorization, is contrasted with the indigenous value-system that relies on storytelling and interconnectedness. They wait a long time before the elderly hunter agrees to start the research trip. At a critical point along the way, Tulugaq warns about a storm, sees the grieving bear and entreats Jean to go back to the skidoo, but the latter fails to do so and falls through the ice.

The seventh scene of the second act details Jean's encounter with Nuliajuk, the spirit who introduces herself as "Inuit Goddess. Of the Ocean. / And the Underworld" (*Sila* 83). Nuliajuk, "played by an oversized puppet or by the same actress who plays Daughter" (*Sila* 3), tells her story in which she identifies as a girl who, when in danger, was not protected by her father (*Sila* 84–86). Even if he recognizes the psychological and physical states he is experiencing, "cognitive disturbances" and "cerebral hypoxia" (*Sila* 84), Jean leaves rationality aside and promises Nuliajuk to protect her (*Sila* 86), thus making an implicit commitment to work for the conservation of the Arctic Ocean as long as he is alive. Once he has been saved and taken to Tulugaq's house, the scientist finds out the whole story centered on Nuliajuk from his host. First opposing the indigenous way of interpreting his underworld experience, Jean ends up trying to follow the hunter's advice, that of combing Nuliajuk's hair to appease her, by doing the same gesture for Veronica, the woman affected by a traumatic experience in his proximity. If seen as a character in the process of "decolonizing his own mind" (May 261), Jean demonstrates that he has advanced considerably on this path.

In a very moving scene (*Sila* 95–99), the antepenultimate in the play, Jean, whose friendship to Leanna was established from the very beginning, visits her house, talks about the loss in his own life[35], empathizes with Veronica, combs the latter's hair, thus unlocking her tears and triggering the much-needed release that could be the first step towards her recapturing her voice. Following a

---

35 He gives the painful details of how his unborn son died (*Sila* 97), blaming himself for not protecting his wife and allowing himself to appear vulnerable in front of his friends, thus reaching out towards thorough healing and a sense of communality beyond any barriers that gender and ethnicity might pose.

circular movement, the end of play presents Veronica performing spoken-word poetry again. Some of her most memorable lines, also by Taqralik Partridge, are:

> maybe we walk the shadowed valley
> or climb up on ridge
> high up on bridge
> inclined to jump – but wait
> while yet we breathe
> it's never too late. (*Sila* 104)

Veronica's return to her art announces her relying on fighting for her indigenous community and for the Arctic as a form of coping with her personal tragedy. The lines she utters further reveal her decision to do her small part in protecting life.

Committed to representing the intricacies of the Arctic territory belonging to Canada, Bilodeau interweaves the scenes centered on Leanna, Veronica, Tulugaq and Jean with scenes featuring two officers from the Canadian Coast Guard Marine Communications and Traffic Services, Thomas and Raphaël. The climax of this strand in the play is reached when, forced to put aside the emergencies in their own private lives, the two do their best to save the crew of a German exploration ship caught in a storm, once again proving the value of human solidarity. But long before this rescue attempt, a conversation between Jean and Thomas, who used to be close friends, reveals the geopolitical importance of the area. Thomas explains to Jean how things stand, clarifying the need for a stronger Canadian presence in the Arctic: "Taking the lead in exploiting our resources is one way to assert sovereignty. Having you, a CANADIAN and one of our most prominent scientists, doing research is another. It shows that we're interested. It shows that we care" (*Sila* 16).

Even if uttered in the twenty-first century, this speech echoes some of the nation-oriented ideals at the heart of the late nineteenth century Norwegian exploration on which Bilodeau's next play centers. Another similarity is the fact that both *Sila* and *Forward* portray the type of scientist that is ensnared by the object of his study. Jean, the Québécois scientist, has a near-death experience in the arms of an ocean goddess, while Fridtjof, the Norwegian scientist, endangers his life when enchanted by ice, both plays requesting their audiences to suspend disbelief and accept the insertion of certain elements that defy the conventions of realist drama. Last but not least, both plays resort to fragmentariness, inviting the audience's active involvement in making sense of the multiple storylines and numerous characters.

Forward – Nansen as the focal point of Norway's relationship with the Arctic

First produced in 2016[36], *Forward* goes back in time to Fridtjof Nansen's *Fram*[37] expedition (1893–1896), an event which is presented as "collective trauma" (Chaudhuri xvi) for an entire society. The word "fram" means "forward" in Norwegian, which explains the title Bilodeau chooses. The play (re-)imagines Nansen's obsession to reach the North Pole and the intimate relationship he develops with Ice, who becomes a central character in this play, while shedding light on the consequences Nansen's journey of discovery had for the next generations of Norwegians. Scenes focusing on the expedition alternate with scenes whose protagonists have been affected by it in various ways.

The play opens with *Fram*'s crew making preparations to leave for the North Pole. Their sea shanty is followed by a song sung by Ice, who is projected as a "magnetic and fierce / enticing and dangerous" presence (*Forward* 10). Her lines, repeated several times, are foreboding rather than reassuring:

> COME TO ME
> I WILL HOLD YOU
> FEED YOU
> SHELTER YOU. (*Forward* 10)

Ice seems to lure Nansen away from his family. The dialogue between Nansen and his wife, Eva, reveals that she opposes her husband's plan, as she is frightened by the prospect of being left behind, even while she is pregnant. She asks him to postpone the trip, but Nansen's answer underlines his hope to obtain an important victory for his country as soon as possible:

> In a few years it'll be too late
> Someone will have beaten us to it
> Norway has few resources
> and little political power
> This is our chance
> to gain independence from Sweden
> and win a seat at the world's table. (*Forward* 13)

---

36 The printed version of the play reveals that "A workshop production of an earlier version of *Forward* was presented by Kansas State University in Manhattan, Kansas, on February 4 to 14, 2016" (2).

37 Fram is the name of the ship that has an entire museum dedicated to it on the peninsula of Bygdøy in Oslo. The museum website introduces it as "the most famous wooden polar vessel in the world" ("Museum History").

He justifies his decision by explaining the political and diplomatic consequences of his much-desired achievement. Presenting his mission as instrumental in securing Norwegian self-determination and international visibility does not make Eva feel better, his ill-timed expedition already proving to have disastrous effects on his personal life. His resorting to scientific arguments and revolutionary strategies – "We'll freeze *Fram* into the pack ice / and let the ice carry us" (*Forward* 13) – is not convincing either. Not even his appeal to their bond as husband and wife who support each other or the need for trust within the couple manage to pacify Eva, since she also needs her husband's support more than ever, his presence and involvement in the raising of their child being more important to her than all the political and scientific matters he invokes and wishes to jeopardize his life for. As a matter of fact, her will and needs as wife and mother are viewed as secondary and dismissed. She is left inconsolable. Her loneliness, desolation and longing are confirmed the next time she speaks, in the play's fifth scene set a year later, in 1894, when, during an imaginary conversation with her husband, she confesses: "I have been so miserable since you left" (*Forward* 29). Being an opera singer, Eva imbues the play with her voice, which cannot surpass the singing that Nansen hears from the personified ice, literally and metaphorically. Songs are interpreted on stage by the actress playing Ice, a note from the playwright specifying: "Original music for ICE was composed by Norwegian electropop singer-songwriter Aggie Peterson of the band Frost" (*Forward* 5). This note shows the playwright's preoccupation for collaborating with Norwegian artists, for using culturally adequate material on stage, and for finding the best strategies of enhancing the play's impact upon the audience.

With the third scene, the play follows Nansen and his crew on board the *Fram*, the power dynamics between the explorer, the ship's captain, Sverdrup, and its stoker, Johansen, as well as the difficulties they face as they advance northwards. Hope and trust that the journey will go according to plan are juxtaposed to doubt and panic. At the center of their preoccupations remains Ice, whom they relate to constantly. To Sverdrup's apprehensive statement, "She's not known for being kind," Nansen responds with:

> But I bet you she can be seduced
> Like any woman
> With the right words
> the right gestures
> the right promises. (*Forward* 22)

Here Bilodeau exposes the explorer's patriarchal mindset; he assimilates the Arctic territory he is set to overpower with the feminine, evincing a conqueror mentality often criticized by the ecofeminist theoretical approach, which takes issue with "the *andro*centric dualism man/woman" being superimposed on "the anthropocentric dualism humanity/nature" (Garrard 23). The playwright disapproves of the scientist's approach to his mission by having his indecent embrace of Ice end with a life-threatening condition. At the end of the tenth scene, Nansen succumbs to the charms of Ice, after hearing the last two stanzas of her song:

> BUT BABY THERE'S NOTHING
> LIKE THE RAWNESS OF MY KISSES
> OH BABY THERE'S NOTHING
> LIKE THE FIERCENESS OF MY LOVE
>
> OH BABY THERE'S NOTHING
> LIKE THE GREATNESS OF MY RICHES
> OH BABY THERE'S NOTHING
> LIKE THE FIERCENESS OF MY LOVE. (*Forward* 54)

The outcome of Nansen's love affair with Ice, the natural element that he is ready to give his life to, is presented at the beginning of the thirteenth scene when Johansen and the crew "carry an unconscious Nansen back to the ship" (*Forward* 60) and try to revive him.

In the long run, he will get close to the North Pole and he will return home to tell his countrymen of this achievement:

> Friends
> I am proud to say that
> at 86 degrees 13.6 minutes north latitude
> Norway now holds the record
> for getting closest to the North Pole. (*Forward* 108)

His triumphant speech in the last scene of the play foregrounds the type of information that is normally recorded by official history. At the same time the interpretations that might undermine this jubilant version of the events are silenced. A sailor in the crowd that welcomes *Fram* grumbles about Nansen's leaving the ship in order to be able to advance farther and thus not doing "his duty as a naval commander" (*Forward* 106), but this ordinary man's dissent is swept away by the crowd's loud enthusiasm surrounding the ship's return. It is the play's merit to have emphasized several elements that lead to alternative readings of Nansen's personality and actions.

The narrative focused on Nansen's advancement towards the North Pole is not continuous and linear; the playwright alternates the scenes featuring the events of the 1893–1896 expedition with moments in the history of Norway since then, reflecting on a multitude of representative lives. The play is built through the accumulation of numerous experiences that give insight into a Norwegian ethos, or in Tale Naess's words, "through a chorus of voices, in a polyphonic bric-a-brac" (xxiii). Using reverse chronological order, Bilodeau acquaints the audience with two partners on the verge of welcoming a newborn (2013), researchers discussing Nansen's mistake (2007), policy experts revisiting Nansen's office (2000), skiers reflecting on the ice decline (1995), a couple contemplating the prospect of buying Nansen's house (1987), an old woman's warning about the ice melting (1975), farmers wondering about climate change (1970), men on an oil rig the moment oil starts gushing out of the North Sea (1969), a family being relocated to the city (1956), one of Nansen's colleagues at the League of Nations approaching his second wife to support the opening of a museum for Fram (1930), and children on a potato field looking into the future (1907), among others.

Bilodeau clearly destabilizes the monolithic story of a national hero since the vignettes that make up the play subvert both Nansen's portrayal as champion and his celebratory legacy. She seems to make every effort of presenting a human being with qualities, failings and dreams in the episodes in which he appears and interacts with his contemporaries. Further on, the play offers various viewpoints on the contribution he and his crew had in shaping Norway's destiny through the twentieth and twenty-first centuries, the fragmentary structure forcing the audience to reassemble the whole picture rather than remember one perspective, no matter how convincing it might prove to be.

The scene set in 2013 capitalizes on the ways climate crisis anxiety influences daily life and brings conflict within the family. More specifically the audience witnesses a pregnant woman, Kristina, who is seized by crying while watching a ship sailing on TV. One would expect that such footage would calm her at a time when she needs peace and tranquility, but the images of fjords trigger a set of associations that stir in her despair and fear for the future:

> It's over!
> We passed the threshold
> We can't stabilize the climate anymore
> We can't stabilize anything anymore
> The planet is going to DIE
> And what do I do? (*Forward* 17)

The sense of vulnerability is interwoven with the sense of revolt. Individual action cannot determine a reversal of global warming, but it can at least try to contribute to measures that would make life for the next generation bearable. In this episode Kristina is the spokesperson for those who consider Nansen's expedition to the Arctic, supposedly "a happy event" (*Forward* 19) in its time, the starting point of unfortunate consequences for the environment. The discovery of oil in the area that he opened for exploration brought about destruction in the name of economic development. In her state of heightened sensitivity, Kristina accuses her partner, Hanne, of playing her own part in the imminent catastrophe: "You work for the fossil-fuel industry!" (*Forward* 18). The latter defends herself by explaining that she is a mathematician, doing her job (*Forward* 18). Moreover, when positioning herself as just another cog in the wheel and concluding that "all we ever do / is our flawed best" (*Forward* 20), Hanne struggles to identify an acceptable answer that would not just save their relationship, but would help Kristina cope with her angst. However, the playwright does not indicate acceptance and reconciliation at the end of the scene since there is no reassurance for those who really think about the scientific evidence available and care about their children's future; Kristina leaves the room with the following battle cry: "That's not enough / Our flawed best is not enough anymore" (*Forward* 20). The visual and auditory impact of a soon-to-be mother's entreaty should be decisive. This maternal urge to support the environment is meant to affect audiences directly and it is reiterated in various forms throughout *Forward*, just as it was in the previous play, *Sila*. At the same time it demonstrates the political overtones of Bilodeau's endeavor in unequivocal terms.

One of the most memorable episodes selected to represent key slices of the Norwegian twentieth century is the one set in 1956. By focusing on one emblematic family, the play attempts to capture the relocation process, a large scale phenomenon that took place in Norway in the 1950s and 1960s. The appeal of urban comfort and of "New beginnings / New opportunities" (*Forward* 76) might be convincing for Ingvar, but not for Kaja, his wife, who mourns the boat that symbolizes her family's livelihood: "My father promised my mother / that this boat would feed her children / and her children's children" (*Forward* 79). Witnessing the sinking of the boat[38] is a traumatic moment for Kaja; her sadness overshadows her husband's optimism and hopes for a better future. Her depression seems to be seeping through the walls of the new house they will most probably inhabit in the city. She defines what they are experiencing as "forced relocation" and condemns the authorities who have initiated it, destroying the traditional way of life in rural Norway: "The government needs labour to build a modern society / so it's luring us with money / to make us all move to the city" (*Forward* 78). With

---

38 The logic behind the need to destroy the boat is explained in the play: "For as long as we have the boat / the government can't sell our quota / to another fisherman / and we can't get the full subsidy" (*Forward* 99).

lucidity and articulateness, she exposes the dangers of industrialization to the detriment of the environment and sustainability. People's caring relation with their land through farming and fishing is jeopardized along with a whole set of practices that have informed the Norwegian sense of self through the centuries. At the end of the scene Kaja is stuck to the place she feels a deep connection to, while her husband and daughter move forward, embracing the dream of "a better country" (*Forward* 79) and a more financially secure life. This family's division, brought about by contrastive attitudes towards relocation, undoubtedly mirrors the Norwegian society's struggle in those years.

In her heartfelt introduction to the play, Norwegian playwright Tale Naess reflects on her own mother's story, which she sees as "the story of modern Norway" (xxv) and which is taken up by Bilodeau's *Forward*. Naess opens her piece with a description of her mother's childhood in the countryside, capitalizing on the lasting effects of her moving to the city (xix). Towards the conclusion, Naess confesses:

> My mother became a writer; she also runs a museum, and she advocates for the preservation of the coastal landscape, its traditional crafts, and its oral history. Her whole life she has been motivated by her homesickness. Not only for the place she was robbed of, but for the life that once was. In many ways, her story now lives in me, and it is present in Chantal Bilodeau's play. (xxv)

By endorsing Bilodeau's project, the Norwegian artist is validating the Canadian author's respectful treatment of the subject matter.

Conclusion

Based on mutually beneficial exchanges between the playwright and the cultural groups portrayed, *Sila* and *Forward* are united by Bilodeau's programmatic ecodramaturgical focus, despite the two works' apparently disjunctive subjects. Committed to representing the intricacies of intersectionality still pervasive nowadays, the Arctic Cycle's initial plays are proven to share technical and thematic similarities. On the one hand, in both cases Bilodeau resorts to fragmentariness and personification of non-human elements, the difficulty of which is successfully surpassed on stage, as shown by the analysis above. On the other hand, without overlooking the need for cultural specificity, Bilodeau establishes favorite themes, shared by the two works discussed: interconnectedness between all life forms; male scientists' obsession with the object of study and subsequent forms of literal and figurative entanglement; feminine (especially maternal) responses in the face of filial loss and environmental crises; empowerment through art, be it music, poetry, sculpture or theater. Clearly innovative, even revolutionary, the Arctic Cycle has ambitious scope, making audiences already acquainted with it look forward to the upcoming plays that will complete the series.

References

"Arctic Cycle." *Chantal Bilodeau. About.* https://www.cbilodeau.com/arctic-cycle. Web.

Arons, Wendy and Theresa J. May. "Ecodramaturgy in/and Contemporary Women's Playwriting." *Contemporary Women Playwrights into the Twenty-First Century.* Eds. Penny Farfan and Lesley Ferris. Basingstoke, Hampshire and New York: Palgrave Macmillan, 2013. 181–196. Print.

Arons, Wendy and Theresa J. May. "Introduction." *Readings in Performance and Ecology.* Eds. Wendy Arons and Theresa J. May. Basingstoke, Hampshire and New York: Palgrave Macmillan, 2012. 1–10. Print.

Bilodeau, Chantal. *Forward.* Vancouver: Talonbooks, 2017. Print.

Bilodeau, Chantal. *Sila.* Vancouver: Talonbooks, 2015. Print.

Chaudhuri, Una. "Foreword. Earthly Love: Biophilic Passion in the Time of Climate Chaos." *Forward.* Vancouver: Talon Books, 2017. xv–xvii. Print.

Fragkou, Marissia. *Ecologies of Precarity in Twenty-First Century Theatre. Politics, Affect, Responsibility.* London and New York: Methuen Drama, 2019. Print.

Garrard, Greg. *Ecocriticism.* London and New York: Routledge, 2012. Print.

Glotfelty, Cheryll. "Introduction: Literary Studies in an Age of Environmental Crisis." *The Ecocriticism Reader.* Eds. Cheryll Glotfelty and Harold Fromm. Athens and London: U. of Georgia P, 1996. xv–xxxvii. Print.

May, Theresa J. *Earth Matters on Stage. Ecology and Environment in American Theater.* London and New York: Routledge, 2021. Print.

"Museum History." *Fram – The Polar Exploration Museum.* https://frammuseum.no/the-museum/museum-history/. Web.

Naess, Tale. "Introduction. A Few Words from the Dramaturg." *Forward.* Vancouver: Talonbooks, 2017. xix–xxv. Print.

Nixon, Rob. *Slow Violence and the Environmentalism of the Poor.* Cambridge, MA and London, England: Harvard UP, 2011. Print.

Sandberg-Zakian, Megan. "Introduction." *Sila.* Vancouver: Talonbooks, 2015. i–v. Print.

Part III: Ecocritical Readings

Anouk Aerni

"Time kept them there and time would let them leave": An Ecocritical (Re) Consideration of James Webb's *Fields of Fire*

ABSTRACT: The global ecological crisis of today serves as a reminder of the ways in which the anthropocentrism at the root of Western culture has facilitated the establishment of a harmful attitude towards nature. In order to transcend the limits of our understanding, a renegotiation of the nature-human relationship based on a more eco- rather than anthropocentric approach to nature, humans and their relation is crucial. As will be exemplified in the following, (re)reading literature from an ecocritical standpoint yields important insights and offers numerous touchpoints for the establishment of a more ecocentric worldview that accounts for the interconnectedness, interaction and reciprocity that characterize the nature-human relationship.

Introduction

In light of the ecological crisis, the inseparability of nature and culture is becoming increasingly more evident. Thus, a renegotiation of the predominant conviction in Western culture that posits humans as essentially different and separable from nature is inevitable. As Dipesh Chakrabarty aptly states, "humans […] have become a geological agent on the planet" (209) making it imperative to reconsider the traditional dualistic conception of nature and humans to account for their intricate connection and reciprocity. One of the ways in which the establishment of a more eco- rather than anthropocentric conception can be aided is by (re-)reading literature through an ecocritical lens and shifting the focus away from the human towards nature.

In that sense, the following ecocritical analysis will demonstrate the ways in which James Webb's *Fields of Fire* (1978) challenges the absolutism of the binary opposition of humans and nature and, instead, suggests a relation based on interconnection and interaction. Through its use of particular themes and devices – including but not limited to the decentering of the human, fragmentation, anthropomorphisms and the omnipresence of death – the novel rejects anthropocentrism as well as the ideas of human exceptionalism and anthropological difference.

Inspired by, and drawing on, Deep Ecology and ANT (Actor-Network-Theory)[39], humans and nature will be understood as inseparable, reciprocally influential and intricately connected yet to a certain degree distinguishable entities. This understanding is loosely based on James Lovelock's Gaia Theory, in which he defines Gaia, our Earth System, as a superorganism, "something that includes individual organisms but exists as a recognizable entity" (133), emphasizing the pivotal fact that everything within the Earth System is interconnected. Thus, this paper will show that reading postwar literature ecocritically may generate vital contributions to an ecologically motivated reformulation of the complex relationship between nature and humans.

First, the ways in which the novel's overall fragmented structure and the fragmentation of its characters decenter the human and thereby challenge the fundamental axioms of Western anthropocentrism will be presented. Subsequently, focus will be shifted towards the tension between empirical and experiential time, how this further distances the human characters from their cultural background and how anthropomorphisms and amalgamations of the human and the nonhuman contribute to an establishment of a more nuanced understanding of the nature-human relationship. Finally, I will close with the culmination of these themes and devices into *memento naturae* – a term that will be further explained in section 4 – and how all of these aspects work together to paint nature, humans and their relationship as one not defined by difference but rather similarities, interconnection and interaction.

The Narrative Structure of War

Fields of Fire is a Vietnam War combat novel written by former Senator of Virginia and "soldier-author" James Webb (Herzog 5). While it is a work of fiction, Webb's postmodernist realist novel is shaped by his own experiences in Vietnam (Herzog 6, 110). Webb is "a highly decorated Vietnam veteran (Navy Cross, Silver Star, and two Purple Hearts) […] from a military family" (Herzog 110–111) who served as Captain in the US Marine Corps before he was retired

---

39 In line with deep ecology philosopher Arne Naess, nature's value is understood as intrinsic, i.e. as "independent of the usefulness of the non-human world for human purposes" (Naess 37) and in accordance with Bruno Latour and ANT, everything – whether human or nonhuman – will be considered as possessing agency (cf. Latour 71). As a result, nature cannot be reduced to a mere background upon which humans act or to a metaphor or dramatization of human affairs. Rather, it will be considered an agent in its own right.

for medical reasons following an injury. *Fields of Fire* was his debut novel and is especially known for its "combat realism" (Herzog 108).

The novel presents the lives of multiple soldiers – five central main focal characters and several less elaborately outlined minor focal characters – in third-person omniscient narration. There are not always clear markers to distinguish between the narrator and the characters' voices, demanding an attentive reader to discern the changes in register. Linda Hutcheon identifies this phenomenon as a characteristic of postmodern literature: "The frequent switching between first and third person complicates the rooting of subjectivity in language, in that it both inscribes and destabilizes it at the same time" (85). Though Hutcheon's observation is not based on *Fields of Fire*, this phenomenon is decipherable in the novel, as the often unmarked changes in register result in or rather add onto the fragmentation of the characters' respective individual subjectivity. This typically postmodern decentering of the self and a figure's individual identity is reflected within *Fields of Fire* as a result not only of the postmodern times but of the war and postwar genre or rather the war as subject matter.

As a result, *Fields of Fire* deindividualizes and dethrones its human characters by exposing the individuals' instability or lack of unity and their insignificance in the grander scheme of war. The characters' fragmentation and deindividualization is reflected in the way in which their voices blend together and, especially, in "the ritual renaming by which the soldiers are reborn into the unit that forms their collective self" (Carton 302). With that, Carton refers to the fact that they are ascribed nicknames upon their arrival in Vietnam, symbolically stripping them of their pre-war identity. From an ecocritical perspective, this depersonalization and deindividualization is especially interesting as it leads to sentiments of no longer being human: "You spend a month in the bush and you're not even a marine anymore. Hell. You're not even a goddamn person." (*Fields of Fire* 75). This is also reflected in the nicknames of some of the soldiers, which include *Cat Man*, *Snake*, *Rabbit*, *Wolf Man* and *Stork* (cf. *Fields of Fire*), inherently carrying the suggestion that the soldiers are no longer human in the dualistic sense of the term. Thus, the culmination of a variety of factors leading to the soldiers' loss of sense of self is the loss of their humanity.

In addition to the characters' fragmentation, the novel as a whole appears decentered and its disunity is reflective of the way that postmodern works in general "challenge narrative singularity and unity in the name of multiplicity and disparity" (Hutcheon 90). The novel's overall structure – or, as Pilar Marin states in her article "Dominant Themes in the Literature of the Vietnam War," the lack thereof (cf. 9) – provides a challenge for the reader. The characters can be read as emblematic of this overall structural disunity. For example, even though there are clear temporal markers and references, the chronology of the overall narrative remains rather unstructured:

> The structure of the novel is not very coherent. In theory, it alternates chapters which give us, as in a flashback, the biography of characters, with others dealing with their experiences in Vietnam. But there doesn't seem to be a real plan to the book. [...] Fields of Fire is at its best in dealing with combat in the jungle, with the way guerilla war affects the psychology of the soldiers and with how these men feel about the war. (Marin 9)

Marin's assessment of the novel identifies the lack of structure as an issue for the reader. The flashbacks cause temporal disorientation for the reader and the large cast of characters further complicates the readers' navigation of the novel as characters are named, renamed, referred to alternately by their respective given names and nicknames, added and removed. Yet, this lack of structural unity does not have to be read as being at odds with the main strength of the novel, as identified by Marin as its realistic depiction of combat. Rather, it can be seen as a reflection of the soldiers' experience of the war:

> The chaos and impermanence of guerilla war will psychologically affect them [the soldiers] and it will be reflected time and again in the structure of the novels.
>
> [...] It is guerilla war that gives shape to Vietnam novels. They are formed by a series of incidents in which the climax is survival or death [...]. (Marin 8–9)

And it is precisely this characteristic of guerilla war that distinguished the Vietnam War from earlier conflicts, such as World War I or World War II and which "influences the structure of the novels [...] in denying them a solid storyline" (Marin 9). Thus, guerilla warfare is reflected in *Fields of Fire* not only in the ways in which it affects the soldiers but in how it affects the narration of the war, resulting in a lack of structural unity.

A similar assessment by Tobey C. Herzog of the author's and the novel's strengths as lying in "the author's descriptions of combat – language, events, emotions, conflicts, and tragedies inherent in the physical and psychological realities of war" (110) shows where the focus of the narrative lies, namely in "accurately describ[ing] the day-to-day life of combat soldiers" (Herzog 110). Given the fact that guerilla warfare is anything but clearly structured and ordered but rather a staccato sequence of chaotic and often deeply disturbing events, a clearly structured, unified and linear novel would not have the ability to aptly reflect the Vietnam War. Thus, the lack of structure in *Fields of Fire* should not be dismissed as a stylistic shortcoming but as one of its strengths, at least from a point of view that considers the novel a realist depiction of combat in Vietnam. In that sense, the novel is both a postmodernist and a realist rendition of the Vietnam War since fragmentation is a stylistic feature of postmodernism and in war narratives it is also realistic.

Therefore, while Marin acknowledges both the genre-specific structural idiosyncrasies of Vietnam War novels and the lack of structural unity in *Fields of Fire*, more emphasis could have been put on the connection between the two. By linking the lack of structure – which is especially noticeable towards the end of the novel (cf. Marin 9) – to the circumstantial chaos, confusion and alienation evoked by the context of war, *Fields of Fire* can be said to reflect the general disorder of war and the particularities of the Vietnam War both formally – i.e. structurally and stylistically – and substantially.

Spatiotemporal idiosyncrasies of the Vietnam War experience

The novel's treatment of time and space as experiential rather than fixed constants works together with the specific decentering of individuals in the war – and, in a broader sense, the general decentering of humans – to initiate a renegotiation of anthropocentrism and the predominant Western understanding of humans, nature and their relationship. The characters are rendered incapable of developing a clear sense of self and, as a consequence, the world, as they find themselves in a context in which their premediated understanding is at odds with their experiences.

One of the dominant themes in Vietnam War Literature or war literature more generally is alienation (cf. Limon 6; Herzog 32), removing the soldiers from the society they left behind. Thus, the physical removal resulting in the soldiers' spatial displacement – in the case of the Vietnam War this takes them halfway around the globe – is amplified as it is experienced on a psychological, emotional level as well. The Vietnam War, particularly, evoked a heightened sense of alienation through feelings of extreme separation for the soldiers from the world back home (Marin 8), making it especially difficult for them to reintegrate themselves into society upon their return (Herzog 15).

The spatial and temporal displacement within the context of war is beautifully captured in Lieutenant Hodges' experience of time in *Fields of Fire*:

> Time was everything. Time kept them there, and time would let them leave. [...] Each day counted as one when a man marked it off on his calendar, but a day could be a week, an hour could be a month. An afternoon spent under fire [...] would age them all a year, [...] yet when the patrol came back on the company perimeter it had been a year for them but only an afternoon for the ones who had not been on the patrol. And

when the cruel sun fell below the western mountains, it still had only been a day. (*Fields of Fire* 234)

As can be deduced from this excerpt, time feels different in the war. The warzone is a timeless world. Life in the "bush" is portrayed as somewhat surreal, in that a principle that would otherwise be considered empirical fact, such as the linear continuity of time, is rendered unreliable. There is a tension between, on the one hand, the affirmation of the importance of time, the fact that time determines the duration of their deployment and acts as a marker of progress and, on the other, the rejection of the concept of time as a steady continuum.

Experiential time appears as something volatile rather than a reliable structural element. Thus, the novel's approach to time, as filtered through the soldiers' combat experiences, displays a tension between nature as indifferent and uncontrollable but also of humans as depending on and embedded within time and therefore nature, thus positing humans and nature as intricately connected. This reinforces the common thread of alienation and estrangement that befalls the soldiers and which runs throughout *Fields of Fire*. Moreover, it adds the dimension of helplessness and other-determination common to the majority of modern war and postwar texts (Herzog 31), evoked by the awareness of the fact that in the war, the soldiers are not self-determined individuals but rather other-determined cogwheels in the military machine.

This typical postwar sentiment of powerlessness and loss of agency is evoked by two intricately linked parameters in the novel, both tied to its depiction of the soldiers' experience of time. For one thing, the soldiers are "kept" in Vietnam by their consignment, determined by the military. Therefore, time determines their lives. However, their utter helplessness is increased by the fact that time does not feel like a reliable parameter, as the soldiers' personal experiences of time defy the laws of chronological time. With that, war is portrayed as a space in which the soldiers' experience and knowledge of time do not coincide.

The archetypal sentiments of alienation and isolation, thus, are linked with the loss of free will and agency, as the former, in combination with the soldiers' awareness that they are other-determined and irrelevant as individuals, induces the latter. Hodges' dissatisfaction with this circumstance of being other-determined is further emphasized in the last sentence and the reference to the setting of the sun and the use of the word "cruel" to describe the process. The rising and setting of the sun is that which determines time's measurement, which, in turn, determines their lives.

In addition, the reference to the sun implies awareness of the fact that time is something out of human control, something natural, steady and continuous, which stands in stark contrast to the soldiers' portrayed experience of time as fickle. With this tension, the passage suggests a relation between humans and nature that is not clear-cut. While nature may be indifferent to some degree, as there are processes within nature that cannot be influenced or altered by humans, such as the rising and setting of the sun, the effects of these processes on humans are at the center of this passage, foregrounding the interaction and interconnectedness between the human and the nonhuman. Thus, there is a constant tension between time as a linear structural element beyond human control and time as experiential or perceived as a personal, inconsistent variable. That, in turn, is emblematic of the ways in which the human and nonhuman are still portrayed according to the traditional Western dualism but, upon closer inspection, reveal themselves to be interwoven in ways that cannot be captured or expressed when adhering to the binary model.

The irreconcilability of time as a reliable measure and experiential time as well as the experiential spatial idiosyncrasies of the combat zone is also explicitly reiterated in the following excerpt:

> In a fraction of a second, a whole hour of events occurred. [...] And even, not more than a second or two having passed, being able to observe with a sort of haunted, detached objectivity, his own frantic effort to retreat behind the mounds again, as if the two feet he had to run were a hundred-yard dash, a cross-country mile. (*Fields of Fire* 396–97)

Similar to the way in which Hodges' experience of time in the "bush" is narrated, the ambush Goodrich and the company are caught in is said to unfold differently, experientially than empirically in terms of time and space. In addition, Goodrich's skewed perception is not only temporal but spatial as well. He experiences two feet of distance as "a hundred-yard dash, a cross-country mile," which, notably, are two different measurements as well. A hundred yards is the equivalent of 0.05 miles, a cross-country mile is one mile. The first comparison implies a short sprint with the term "dash" requiring speed, whereas the "cross-country mile" implies a longer run on uneven ground, demanding endurance and surefootedness. Thus, his appraisal of the distance is based on the experienced labor, that is to say on his perception of pace, accuracy and stamina. This again underlines that subjective perception and experience are at a discord with objective or empirical reality in combat, exposing the limits of human understanding as based on cultural constructs.

Passages such as these in combination with the narrator's and characters' reiterations of feeling like they are in an "other world" (*Fields of Fire* 152) further solidify the soldiers' experience of temporal and spatial alienation in the war. According to Marin, this particularly radical sense of geographical separation experienced by the soldiers is emblematic of the Vietnam War experience given that the war took place halfway across the globe (Marin 6) and due to that the soldiers' "feeling of separation was never as strong as in Vietnam" (Marin 8). The world of the Vietnam War is a world apart. It is a world in which individuals have no value, power, agency or control that transcends that of their designated function within the collective. In other words, it is a world in which the soldiers are stripped of many of the features that are traditionally considered uniquely human. The alienation they feel from society brings with it a feeling of isolation from humankind as defined by culture. It removes them from themselves as their individual identities are rendered obsolete. As a result, this negotiation of the relationship between the civilian world and the world of the war and, in connection with that, the relationship of the cultural environment the soldiers once felt a part of and the alien environment they are thrown into facilitates the establishment of the two as divergent and yet connected.

Amalgamations and anthropomorphisms

Ab initio, the novel's engagement with the topic of the nature-human relationship is apparent. Sentences like "[t]hick white smoke rushed from every opening of the pagoda, and mixed with the low rain mist" (*Fields of Fire* 11) from the beginning of the novel or "blood and water mix it's all the same" (*Fields of Fire* 401) from towards the end, continually blur the lines between the human and the nonhuman as their interconnectedness is emphasized, in these cases especially overtly through the use of the word "mixed." However, one of the most self-evident ways in which humans and nonhumans are put on equal footing is through the ascription of agency to nature via anthropomorphisms.

While anthropomorphisms as such are not *inherently* ecocentric as they may also result in the devaluation of nature, they often function as enablers of nature's agency. Anthropomorphizing means attributing features that are typically understood as human to a nonhuman entity. As Bruno Latour states, "*anything* that does modify a state of affairs by making a difference is an actor" (71) and thus, anthropomorphisms infuse that which is traditionally considered passive with life, making it an actor. For example, choosing the active verb form to describe activities performed by a natural entity highlight the entity's agency, autonomy and status as an actor, making the thereby emergent subject equal to a human agent, as it is illustrated by the following passages from the first chapter of *Fields of Fire*:

> A gust of wind swooped down from the amber mist of sky and chased him, rattling trash. […] His eyes scanned the building quickly and his narrow shoulders raised against the

biting wind, but otherwise there was no reaction from him. [...] He climbed the outside steps [...] finally being chased inside by another gust of wind. (*Fields of Fire* 8) [...] Then he walked quickly, half-jogging through the streets. It was late and bitter cold. The air attacked him and he shivered nakedly in its crispness. (*Fields of Fire* 20)

These excerpts are an example of nature being ascribed agency simply by being given a voice or rather a space in which to act, as the main antagonist of the human focal character is not human. The wind *swoops*, *chases*, *rattles* and *bites*, the air *attacks*. Especially in the case of the last four of these verbs, which are transitive, nature is posited as an active adversary. Reducing the nonhuman to mere background or setting would, in this case, be a gross oversimplification of a much more complex interplay of entities since the active verb form quite overtly attributes agency to the nonhuman. The only action on part of the human, in contrast, consists of the movements of isolated body parts, which diminishes the human's importance as an individual, much like the fragmentation at the center of the preceding section. The nonhuman performs the actions that necessitate all other (re)actions within the passages. With that, these excerpts foreground the ways in which human actions are, at times, mere reactions within an active environment, clearly undermining the human primacy implied within the nature-human dualism.

Via anthropomorphisms such as this, the text continuously equates the human and the nonhuman, exposing features and actions traditionally viewed as strictly human capacities as capabilities of both the human and the nonhuman, rejecting anthropological difference and human exceptionalism. The main effect of these equations of the human and the nonhuman via the ascribing of agency to nature and the therein contained rejection of the idea of anthropological difference and human exceptionalism is the blurring of the artificial boundary between humans and nature as they are put on equal footing as acting agencies. The result is a debunking of the human-nature dualism in favor of an alternate conception of nature and humanity that values both in their own right.

In addition, amalgamations of the human and the nonhuman, as encountered in the mixing of the two mentioned in the beginning of this chapter, strengthen the sense of interconnectedness. The following excerpt further solidifies this transcendence of the dualism:

> The Americans, true to the "temporary" nature of their commitment, erected sandbag bunkers that decayed, sagging at the seams and finally bursting, oozing back into the earth, each monsoon season, and had to be continually replaced. (*Fields of Fire* 210)

This excerpt, though it represents a theme that recurs throughout postbellum literature of different literary periods, is especially significant in the context of the Vietnam War. The allusion to the fact that the American soldiers are only in Vietnam for a limited amount of time is not only true because their stay was determined by the duration of the war but rings true for every soldier individually because, as was "first introduced in Korea, but greatly expanded in Vietnam, American soldiers served limited tours" (Herzog 53). This time limit and the associated transient nature of the soldiers' stay in Vietnam is then juxtaposed with the sandbags. Their likeness is evoked by the fact that just as the soldiers' stay is temporally limited, so is that of the sandbags, which continually disintegrate and warrant constant rebuilding. With that, this excerpt is also a testament to the lasting effects and alterations caused by warfare within the environment.

The cyclical characteristic of the action of placing and replacing the sandbags plays into the notion that the war takes place in a world that is decidedly different from the soldiers' US home in terms of climate, alluding to the temporal idiosyncrasy of said world, which is determined by nature. The construction of the bunkers is a reaction to the weather conditions and therefore a reaction and direct interaction with the natural world through the alteration of the geological conditions of that space via the displacement of the sand.

Moreover, while the coming and going of the monsoon is the determining temporal factor, the military and therefore the cultural objective of furthering the Marines' war efforts is an equally determining factor. With both in mind, the lives of these soldiers are once again posited as other-determined, by nature and culture respectively. This reiterates the idiosyncrasies of the war environment and simultaneously allows for the emergence of similarity between human and nonhuman temporal contingency. The impermanence of the soldiers' commitment is paralleled with the changeability of nature's seasons. By paralleling the human and the non-human in this way and by referring to the ways in which the sandbags "ooze back into the earth," the artificial boundary between the two is washed away along with the constructed sandbag boundary.

As Bronwyn Leebaw notes, "[e]nvironmental devastation is not only a by-product of war and militarism, but has also been implemented as a military strategy since ancient times" (770). The heightened interaction with nature during the war, either deliberately in the case of ecological warfare[40] or as a

---

[40] Especially in the case of the Vietnam War, it is important to consider the effects of ecological warfare not only from a military but also from an environmental point of view: "The war in Vietnam saw the first full-scale use of herbicides in warfare [...] aimed at the defoliation of mangroves and forest, and destruction of crops and their distribution so as to remove aerial cover and food supplies" (Palmer 172). The

byproduct of combat within the natural environment, makes this postwar narrative a paradigmatic example of the ways in which postbellum literature in general offers numerous touchpoints for ecocritical approaches. By foregrounding nature-human interaction as an inevitable aspect of warfare and by pitting the cultural sphere of the civilian world against the war environment, *Fields of Fire* is part of the discourse on the nature-human relationship and the negative influence of anthropocentrism on human attitudes and, particularly, behavior towards and within the natural environment.

Memento Naturae

Thus, anthropomorphisms and amalgamations work in a similar way to *memento mori* – i.e. death imagery – within the novel. Death is omnipresent in war literature in general and aside from functioning as a recurring reminder of human mortality as *memento mori*, literary images of death often also function as reminders of humans' embeddedness within nature by foregrounding mortality. That is why – in this context – *memento mori* can more generally be referred to as *memento naturae*, a term created to encompass the multitude of literary images and devices that work at embedding humans within nature. The attribution of language or other capacities and characteristics traditionally conceived as strictly human to nature also works as *memento naturae*, challenging the nature-human dualism as it functions as a reminder of the similarities and interconnectedness of humans and nature and, especially, the embeddedness of humans within nature. Therefore, anthropomorphisms and *memento mori* can both be subsumed under the term *memento naturae*.

One example of a *memento mori* that is also a *memento naturae* in *Fields of Fire* is the death of soldier Big Mac following the setting off of a booby trap:

> He landed where he had stood grinning to them only a half-second before, but now as a scorched, decapitated ash heap that reminded them all of how very close they stayed to death, even on a boring day.
> Pieces of Big Mac pattered on the leaves and grass for several seconds, like gentle rain.
> [...] They had not noticed Phony staggering nearby without direction. Finally he screamed, high-pitched and confused. "DOC! Doc! They got my arm, Doc! They got my arm!" There was nothing underneath his right shoulder [...].
> (*Fields of Fire* 314)

---

environmental effects, especially long-term, were understudied and therefore unanticipated (cf. Oatsvall 427–428).

This scene beautifully illustrates the shift from *memento mori* to *memento naturae*. As Big Mac's death reminds the soldiers of their mortality, Big Mac is diffused into nature "like gentle rain." Moreover, as he disintegrates, fellow soldier Phony is fragmented; both are literal fragmentations of the self, emblematic of the decentering of the individual and the human within the novel. Big Mac's human body dissipates into nature, functioning as a radical symbol for the human embeddedness within nature, as the comparison between the pieces of Big Mac and rain are not only a *memento naturae* but a literal amalgamation of the human and the nonhuman, as the distinction between the two is blurred.

The novel, then, uses *memento mori* to undermine the nature-human dualism by foregrounding their similarities and portraying humans as not apart from but part of nature. Aside from anthropomorphisms and *memento mori*, explicit comparisons of humans to animals or human made structures to nature also infer connections and similarities:

> The bomb crater was old and deep, as natural to the field as would be a tree or a dike. The shards of earth once set loose from far into the earth had melted back into their origins, two years of rain and sunbake have mended the gaping wound, leaving only its memory in the deep pock of a scar. A thin carpet of yellowed sawgrass continued from the field down one side of it, as if an entranceway. (*Fields of Fire* 84)

The bomb crater serves as a reminder of the disruption of the natural landscape but the focus here lies on its "natural" appearance. With this comparison of a bomb crater, which is human-induced, to natural occurrences, which develop or grow over time, the text associates the human and the nonhuman in a way that focuses on their similarities. The second sentence, then, describes nature's recovery, implying the disruptive human force that caused the crater in the first place. The way the "shards of earth" are said to have "melted back into their origins" is evocative of the creation of a magmatic dike. The term "shards of earth" also clearly marks the destructive aspect, since shards are fragments created with the breaking of a thing. In addition, the way the narrator refers to the "gaping wound" in the earth, which alludes to physical destruction, and the "memory" of the impact and the "scar" that remain are reminiscent of the way in which humans respond to and heal from the infliction of physical and/or psychological trauma.

Thus, the passage suggests that the human and the nonhuman are similar in the ways in which they are affected by the war and thereby functions as *memento naturae*. Scenes such as this reveal the soldiers in *Fields of Fire* as at war against people *and* the environment. These types of comparisons or the abundance of equations of nature and war in general, rain and gunfire in particular[41] or humans and animals[42] within the novel, do not make nature a metaphor or dramatization of the human. Rather, they emphasize the fact that the nonhuman is not so different from or less active, less important or less powerful than humans. Anthropomorphisms and comparisons therefore expose how the human and the nonhuman are characterized and described with the same vocabulary and thus ascribed agency equally. Therefore, from an ecocritical perspective, they affirm nature's autonomy and its value as an active agent.

The overt amalgamation of the human and the nonhuman follows in the last sentence, where the narrator refers to the "carpet of yellowed sawgrass" that stretched along the field "as if an entranceway". For one thing, this shows the ways in which human perception is always mediated and influenced by a person's specific cultural background. Seeing the yellow grass as a carpet implies that humans use their limited and limiting cultural understanding to make sense of the world. The "entranceway" suggests the existence of the existence of two sides and the conditional separability of the two. Yet, the implicit possible transcendence of that boundary via the "entranceway" allows the text to oscillate between connecting nature and humans and maintaining their recognizability as separate entities, painting them as coexisting, interconnected, reciprocally influential entities.

Conclusion

The narrative style and structure of James Webb's *Fields of* Fire greatly contribute to the fragmentation, deindividualization and even dehumanization of its characters. As a result, the human is decentered, which enables the text to stage a critique of anthropocentrism and the thereby associated assumptions of human exceptionalism and anthropological difference. Its postmodern rejection of unity of character is heightened by an overall lack of narrative structure and unity, which is the result of the novel's realistic engagement with the subject matter

---

41 Such as in "[r]aindrops popped and sizzled as they pelted the tiny stove in front of him" (*Fields of Fire* 1), "listening to the drops beat on the rubber like small explosions" (*Fields of Fire* 226) or "[r]ain fell steadily, big drops like explosions on the ponchos" (*Fields of Fire* 289), where the choice of verbs evokes a similarity between rain and gunfire.

42 Explicit comparisons of humans and animals shift the focus from the differences between humans and animals to their similarities: "They approached Hodges slowly, singly or in twos and threes, like hesitant, wild animals inspecting their latest zookeeper" (*Fields of Fire* 88).

of guerilla warfare. The decentering of the human and consequential rejection of anthropocentrism in *Fields of Fire* threatens the traditional dualistic conception of humans and nature that privileges humans and posits them as exceptionally different from nature. In doing so, the text exposes the limits of traditional Western or US cultural thinking. By spotlighting the ways in which the soldiers are alienated from themselves, their past and US society via temporal and spatial idiosyncrasies, the novel further reflects the chaos and aftermath of the Vietnam War specifically substantially as well as formally and continuously renegotiates what it means to be human within and in relation to the world.

The soldiers' experience of war as one of displacement, estrangement, isolation and alienation facilitates a shift away from an anthropocentric and towards a more ecocentric conception of the nature-human relationship. The novel's approach to time, as filtered especially through the soldiers' combat experiences, displays a tension between nature as indifferent and uncontrollable but also of humans as embedded within time and nature, thus positing humans as inextricable from nature. In its use of anthropomorphisms and human-nature interaction, the narrative posits nature as an actor, further debunking the anthropocentric assumption of human exceptionalism. Implicit and explicit evocations of the connection between humans and the nonhuman through *memento naturae* as well as the interaction between and amalgamations of the two are further proof of the interconnectedness between nature and humans and the embeddedness of the latter within the former.

The novel's depiction of heightened interaction with nature during war makes this postwar narrative a paradigmatic example of the ways in which postbellum literature in general offers numerous touchpoints for ecocritical (re-)readings. By foregrounding nature-human interaction as an inevitable aspect of warfare and by pitting the cultural sphere of the civilian world against the war environment, *Fields of Fire* is part of the discourse on and critique of the negative influence of anthropocentrism and the nature-human dualism on conceptions of the nature-human relationship.

As the text oscillates in a space of ambiguity – by constantly embedding the human within nature, extracting humans from nature, re-embedding them, contrasting and then comparing and likening humans and the nonhuman to one another – it provides the ideal breeding ground for, and stages a renegotiation of, the nature-human relationship in the general context of war and the particular framework of the Vietnam War. Thereby it offers a reconsideration of the traditional dualism by picking it up, yet subverting it and, as a result, by portraying nature and humans as intricately connected and humans as inextricable from nature.

References

Carton, Evan. "Vietnam and the Limits of Masculinity." *American Literary History* 3.2 (1991): 294–318. Web.

Chakrabarty, Dipesh. "The Climate of History: Four Theses." *Critical Inquiry* 35.2 (2009): 197–222. Web.

Herzog, Tobey C. *Vietnam War Stories: Innocence Lost*. London: Routledge, 2005. Print.

Hutcheon, Linda. (1988) *A Poetics of Postmodernism: History, Theory, Fiction*. New York: Routledge, 2004. Print.

Latour, Bruno. *Reassembling the Social. An Introduction to Actor-Network-Theory*. Oxford: Oxford University Press, 2007. Print.

Leebaw, Bronwyn. "Scorched Earth: Environmental War Crimes and International Justice." *Perspectives on Politics* 12.4 (2014): 770–788. Web.

Lovelock, James. *The Vanishing Face of Gaia: A Final Warning*. London: Penguin Books, 2009. Print.

Limon, John. *Writing After War. American War Fiction from Realism to Postmodernism*. Oxford: Oxford University Press, 1994. Print.

Marin, Pilar. "Dominant Themes in the Literature of the Vietnam War." *Atlantis* 3.1 (1981): 5–15. Web.

Naess, Arne. "The Deep Ecology Movement: Some Philosophical Aspects." *The Selected Works of Arne Naess. Volume X*. Eds. Harold Glasser and Alan Drengson. Dordrecht: Springer, 2005. 33–56. Web.

Oatsvall, Neil S. "Trees versus Lives: Reckoning Military Success and the Ecological Effects of Chemical Defoliation During the Vietnam War." *Environment and History* 19.4 (2013): 427–458. Web.

Palmer, Michael. "The Case of Agent Orange." *Contemporary Southeast Asia* 29.1 (2007): 172–195. Web.

Webb, James. (1978) *Fields of Fire*. London: Panther 1985. Print.

Pauline Boisgerault

"Not the glow of red or green as in picture-book terror wolves, but a dullish, perversely dignified human gold." An Eco-Critical Reading of *Hold the Dark* by William Giraldi

Abstract: This essay offers an eco-critical reading on William Giraldi's 2014 novel *Hold the Dark*. Estranged from their family or community, the novel's characters struggle to regain a sense of place in their environment. They can only hope to come to terms with their loss of bearings by acknowledging their fault towards the many wolves of the novel, both hybrid and animal figures. After a series of murders perpetrated on children, for which the wolves are falsely accused, the (wo)manhunt through the uncanny Alaskan landscape sheds light on its contaminative and transgressive nature. Yet, these transgressions allow the novel to offer a de-territorializing perspective on human and non-human relations, based on a blurring of boundaries and relying on species interdependence.

Introduction

Hold the Dark's moonscape partakes in enclosing both the narrator and the reader around a murder mystery which seems to delineate Keelut as a place of despair, on the verge of disintegrating itself. Yet, the death of young Bailey pinpoints the sense of togetherness which, if it ostracizes the outsiders, nonetheless enables those within Keelut to form a community. Despite its sense of enclosure, the novel also deploys multiple lines of flight that pave the way for an ecocritical reflection on interspecies relations, favoring openness through a rhizomatic viewpoint rather than a treelike and hierarchical one. Far from being a human self-examination only, the novel's narrative and stylistic devices are put forward to expose the very precarious epistemological and ontological grounds on which species and categories are established. In this essay, I focus on the ways in which language and the depiction of space are intertwined and are paramount in tracing out a new relation between humans and non-humans. As such, the purpose of this essay could be summed up in these terms: "It is about humans and how to get outside of the represented human towards the asignified a human, using the tentative virtualisation of what a nonhuman human could be while also thinking about how to be more graciously ethical towards actual nonhuman animal lives" (MacCormack and Gardner 6).

In order to do so, we will rely extensively on Deleuze and Guattari's view of nature and culture, and of reality and language, which works as a network of rhizomes and thus cannot be dichotomized. They posit that a rhizomatic perspective as a model is a concept which enables one to understand social, cultural and political systems as well as natural and ecological ones. As such, human and nonhuman realities and their representations allow for a mutual and interconnected relation which wards off the hierarchical arborescent model which usually prevails over our thought and our connection with non-humans. This model allows for both to transform one another and helps outline an ecocentric reading of the world, which leads to a human decentering and fosters a community based on a sense of collectivity and interrelatedness. Subverting the human predominance at the top of a treelike power system, ecocriticism questions the supremacy of literary representation and the legitimacy of its narrative voice. In *The Environmental Imagination*, Lawrence Buell contends that "ecocentric literary vision may express itself both as a critique of the centrality and even the legitimacy of human assertion and as an ascription of something like human subjectiveness to the nonhuman world" (Buell 143).

In *Hold the Dark*, this subversion shows through a reflection on its own construction via metafictional comments, but also through the blurring of dichotomies, thus throwing off balance "the authority of the superintending consciousness" (Buell 144). *Hold the Dark*'s plot centers on nature writer Russell Core, who sets out on a journey into the oppressive Alaskan moonscape. His venture reflects a personal quest which brings about questions about human nature and its relation with its surroundings.

Core's trip is set in motion after receiving a letter from Medora Slone, in which she implores him to come to Alaska to help find her son's bones after his being presumably killed by wolves. She claims that her son was snatched by a pack of wolves and she calls upon Core's skills as a wolf-expert to help her hunt those she deems responsible for his death. Medora's husband, Vernon, comes back from the war to find out his wife gone and his son dead. He then sets out on his own journey to find her. Core ends up doing exactly what he was summoned to do, that is, to chase the culprit through the wilderness. Medora Slone is likened to a female-wolf who, after killing her progeny, is chased by the police, Core and her husband.

Tracking both Medora and Vernon, the unfolding of the plot revolves around a (wo)manhunt which explores the hunt leitmotiv through the wilderness, which was examined by Richard Slotkin in *Regeneration Through Violence*. The mystery thickens as spaces in the novel become outlets for the violence within the human characters, but also allow other non-human beings to shed light on species interdependence. I would thus argue that, far from being considered as subservient to a reflection over human nature, the wolves are part and parcel of a larger ecocritical pondering over what it means to simply be, which is to be interpreted within an interconnected and inclusive framework. As such, the mystery surrounding Bailey's death unfolds in an unfamiliar landscape to set off the intricacies of interspecies relations. Going beyond epistemological and ontological dichotomies, the multiple layers of meaning unravel an instability that stems from space itself, which, far from being a backdrop only to the events, plays a center-stage role in the unveiling of the generational secret and underlying angst that corrupts the community of Keelut.

Fencing in otherness

Set in the Alaskan moonscape, William Giraldi's 2014 novel is also his first experiment with the American Gothic literary legacy. His essays flourish with references to Southern Gothic authors such as O'Connor, for whom he has a particular literary admiration. In his essay on literary influence, William Giraldi notes: "And there was Flannery O'Connor's 'A Good Man Is Hard to Find,' a ferocious story by an uncommon genius who astonished my twelve-year-old sensibility and astonishes me still" (Giraldi 2018: 424).

The Alaskan space provides a sense of enclosure which stems from its Gothic atmosphere. The novel's careful interweaving of contrasting spaces partakes in the smothering and puzzling effect one gets from reading the first two sections, separated by an asterisk. The reader has no choice but to feel the spatial experience of the freezing cold Alaska after being transported to a scorching desert in the Middle East: "Outside the desert city an urgent wind whisked up sand. Dark mustard gusts passed before a buffed sun and looked like blots of insects sent to swarm. Their vehicle made plumes of tawny dust as they sped up after a pickup rusted red and packed with men" (*Hold the Dark* 16). The vivid yellow and red hues offer a limited palette which helps the reader construe the place as burning hot and intensely bright. The range of warm colors also allows violence to fit in with the color scheme pertaining to the description of deserts, as is suggested by the blood "burst[ing] in the wind as wisps of orange and red" (17).

Yet, by showing how the bright red of blood stands out even in the desert, the internal focalization on Vernon also sets it apart, as something out of place: "Curious how orange, how radiant blood looks beneath a desert noon, in the dull even tinge of its light" (17). The structure of the incipit navigates the two spaces without any transitions, leaving the reader to adjust to new extremes, replacing one enclosed environment with another through a perspective of displacement, which is reinforced through Core's loss of bearings when landing in Alaska: "[He] was lost immediately upon leaving the terminal" and the city he lands in remains unnamed and alien: "odd city" (18). Further emphasizing the sense of the unfamiliar strangeness and the inability to get one's bearings that goes with it in the novel, the city's name is fictional, Keelut, and impossible to find even on a fictional map: "[…] the attendant told him that where he was going could not be found on GPS" (18).

The sense of bewilderment is further developed thanks to a play on enclosing spaces. The novel paradoxically opens with limits and borders, which is telling of its narrative strategies, playing on the very proximity of danger, getting closer and closer: "at the rim of the village," "as she skirted," "from his own doorstep" (11). The plot revolves around a chronotope which is tinged with a haunting remaining presence of the past, as the latter emerges regularly as reminiscences of past illness and death, which contaminate the present. As such, it tallies with Baldick's definition of contemporary Gothic texts which should comprise "a fearful sense of inheritance in time with a claustrophobic sense of enclosure in space, these two dimensions reinforcing one another to produce an impression of sickening descent into disintegration" (Spooner 14).

The uncanny permeates the novel to the point that the unsettling nature of the moonscape unveils a deep angst which seems to contaminate all relationships with what is seen as "otherworldly" or "unnatural," the last term being used several times to depict Vernon as a child. In fact, contamination serves as a leitmotiv to allow the unfamiliar to find its way in the narrative through echoes alluding to a distant plague brought on by the Christian colonizers, and which seems to trigger and spread illness into the present. The resulting deaths of the village's inhabitants are mentioned by the hag, but seem to loom over the landscape and the protagonists, still affected by the contagion and malaise it induced. "What plague had invaded these vast silences?" (*Hold the Dark* 43). The term *plague* puts forward the causes of the angst which permeates the landscape and the protagonists, as it alludes to both contamination and invasion. As such, Core's reminiscences of warm weather seem to have been obliterated by the icy-cold Alaskan weather: "The vivid earth, his memories of fruit breathing hotly in summer fields – all obliterated by this moonscape" (43). The hyphen evokes this process of erasure in Core's memory, and hints at the interwoven relation between memory and sense of self, his sense of time being unsettled because of the omnipresence of sickness: "He looked to the sky but could not tell time from a sun this sick" (43–44). In attributing his own mental state to the sun, the hypallage helps expand and mirror the contaminative remanence.

While he has an in-depth knowledge of wolves and can be seen as a wolf-expert, Core's understanding of his human peers seems to be rather superficial at first. His journey could serve as a reflection on the difficulty to fathom the inner human consciousness since Core remains out of touch with his human counterparts and serves as the main focalizer throughout the novel. Closer to the lycanthropic world, Core struggles to face his inner turmoil until he finally realizes that the culprit for Bailey's death is none other than the boy's own mother.

The novel also conveys a wish to disrupt boundaries and to transgress them. As such, another reading of the passage from the first chapter to the second also leads to a sense of openness. The plot does not restrict itself to its bordered Alaskan space, but seems to merge with the contrasting aforementioned desert, reinforcing the reader's confusion and loss of bearings. Likewise, the boundaries traditionally erected to categorize and separate the human from animals also crumble down to blur the dichotomous vision that seems to divide and pit humans against animals from the onset of the novel.

Infringing the other's territory

At the onset of the novel, the incipit is fraught with hints at the omnipresence of space as a junction or a threshold, by highlighting its in-between state. In doing so, the incipit and the whole novel are replete with multiple spatial transgressions, starting from the first paragraph of the novel: "The wolves came down from the hills and took the children of Keelut. First one child was stolen as he tugged his sled at the rim of the village, another the following week as she skirted the cabins near the ice-choked pond. Now, in the rolling snow whorls of the new winter, a third was dragged from their village, this one from his own doorstep" (11). This paragraph deserves some comments, as it points out the very polysemous nature of language, which serves to blur the meanings and cast a different light on Bailey's death, when we read it retrospectively. Indeed, here, one could say that space and its moonscape reflect the characters' state of mind, in a kind of specular pathetic fallacy. Indeed, one could read the ice-choked pond as an allusion to Bailey's death, strangled by his own mother.

Before I go any further, I would like to point out the etymology of transgression, which is embedded in a crossing of boundaries, whether they be spatial or linguistic, or rely on some moral sense. All types of transgression occur in the novel. The first spatial transgression here foreshadows the following ones, all originating from human hubris, many of which surfacing through language. Far from all being seen as negative, transgressions also enable one to bewilder where the evil spirit really lies in this Alaskan contemporary Gothic tale.

As the reader reaches the end of the passage, the wolves have gone over the hills to finally trespass on the village as if an invisible frontier had been crossed, a moral and spatial one, as the terms suggest: "at the rim," the verb "to skirt," and "his own doorstep" (11). Now, do the verbs used to depict the disappearance of the children rely on understatement to hint at their killing, or is it really an understatement? Since we know retrospectively that Bailey's body remains difficult to find, because of the unremitting snowfalls, could it be, in this case, that these anthropomorphic verbs "take," "steal," "drag" (11) only refer to those exact actions, that could only then be perpetuated by a human being? In the incipit, the verbs could prefigure the wolves' innocence in this snatching of children, but it should be noted that the terms seem to echo the paragraph about Yellowstone, which I have already quoted. Indeed, the she-wolf was then said to have "robbed" the girl, which seems to play down her killing, rather than point at a human origin.

When we look back at the same excerpt, the passage ends on the absence of noise "Noiseless – not a scream, not a howl to give witness" (11), and the subsequent absence of witness may be seen as suggestive of the forging and construction of this very story in which wolves have supposedly attacked children. In other words, the incipit is based on hearsay – the story Medora has told her neighbors – to emphasize Core's own attempt at making sense of the crime. Indeed, a few lines later, the indirect discourse, rarely used in the novel, relates Medora's false story through Core's retrospective focalization and already depicts her as a figure of transgression: "She told her fellow villagers how she had trekked over the hills and across the vale all that evening and night and into the blush of dawn with the rifle across her back and a ten-inch knife strapped to her thigh" (11).

If the novel sheds light on the difficulty to delve into the human mind, it also subverts the dichotomous boundaries that usually delineate and set humans apart from other living beings. I would suggest that this subversion in fact contributes to getting a better understanding of the human mind. Core's propensity to side and empathize with the wolves runs through the novel and opens the way for a more porous connection between all living beings. As such, the novel's writing style challenges the usual dichotomy, questioning the humans' knowledge of their environment and of the other living beings inhabiting it.

One telling example of how Core's empathetic depiction of wolves partakes in this blurring of boundaries lies in his reluctance to shoot a she-wolf after a girl was killed at Yellowstone National Park. Core blames the girl's parents for camping there and men in general for having led wolves to rely on a shrinking habitat. His portrayal of the animal is tinged with sympathy, striving to question the mythical Christian wolf, epitomizing destruction: "Her eyes were golden. Not the glow of red or green as in picture-book terror-wolves, but a dullish, perversely dignified human gold" (13).

Core's perspective portrays the wolf with anthropomorphic features and, because she is endowed with the same "dull" or ordinary traits as humans, the she-wolf's description bespeaks the wish to tone down the mythical and devilish Christian depiction that the wolves are often stuck with: "He imagined they wanted a story woven of truth, not myth, one not tilted by dread" (12). This comment could be seen as one of the few metafictional comments in the novel, which sheds light on the very way the reader expects the novel's story to be woven.

By showing that the she-wolf acted out of hunger, because she was forced to seek food further away from her habitat, Core's prism turns the argument upside down by blaming the killing on human transgression and expansion as his following comment shows: "she dared to intrude not because it was her wish but because it was her need. Who was the transgressor here?" (14). In establishing the campers as the transgressors, the novel is meant to suggest that humans have infringed on the wolf territory, which is an idea that the novel fleshes out through multiple allusions to human transgression.

In the same paragraph, the nature writer mentions both his daughter and the she-wolf, the stream of consciousness conflating the two, which puts forward the porous boundary between the human and the animal form. While dwelling on his reason for accepting to hunt the she-wolf in Yellowstone, his mind strays from his daughter and onto the she-wolf, the personal pronouns "she" and "her" changing the signified as the paragraph slithers on:

> He could not say no. His daughter was the same age as the taken girl and his love for her then felt already like loss. The guilt of a father whose work takes him from home. He and the others, the rangers and the biologists and camouflaged men, tracked the wolf across twenty square miles over the Northern Range, through Lamar Valley. He rode on a borrowed four-wheeler and was in radio contact with the sniping copters he hated. Then he rode across the line into Montana where, alone and sickened, he found her and shot her from forty yards on a cattle farmer's ranch" (*Hold the Dark* 13).

Still showing him harkening back to his past as a wolf-expert here, this paragraph captures the transgressive power of language, which plays an important part in confusing the reader.

One can say that the writing style, peppered with literary tropes, such as the aforementioned hypallage, opens the way for an in-between state of being, which would be both human and animal. It also constantly reminds the reader of the changing nature of both man/woman and wolf, and their unpredictability. Even though the inner focalization allows to display Core's empathy with the she-wolf, the fact remains that he shot her, even if it was because he thought "his own land-borne bullet more respectful than those from rangers impersonating god" (13). Even if he strives to decry the hubris that motivates his fellow hunters, he is not unfailing himself, which makes him all the more complex as a character.

De-territorializing the animal

The moonscape seems to stage the meeting with the other through an exploration of the uncanny. Nicholas Royle details what he means by "the uncanny" as follows:

> The uncanny is ghostly. It is concerned with the strange, weird and mysterious, with a flickering sense (but not conviction) of something supernatural. The uncanny involves feelings of uncertainty, in particular regarding the reality of who one is and what is being experienced. [...] The uncanny is a crisis of the proper: it entails a critical disturbance of what is proper (from the Latin proprius, "own") [...] More specifically, it is a peculiar commingling of the familiar and the unfamiliar. It can take the form of something familiar unexpectedly arising in a strange and unfamiliar context, or of something strange and unfamiliar unexpectedly arising in a familiar context. (Royle 1)

In *Hold the Dark*, space is often viewed through the eyes of disoriented Core, whose relation to the moonscape seems to be clouded by his own deferred relation to the real.

Confounded by Medora's attitude and by the possibility of her being the murderess, the reader also finds him/herself chasing after the clues interspersed from the incipit to point to the culprit and to sketch out the underlying malaise of which the moonscape is a reflection. Whether Keelut is a hospitable place or not, the alien landscape remains elusive: "The hills beyond loomed in protection or else threatened to clamp" (20) or on the next page: "A set of caribou antlers jabbed out from above the door – in welcome or warning, he could not be sure" (21). Estranged from his daughter whom he has not seen for years, estranged from his wife who cannot speak due to a stroke, the descriptions through his inner focalization bespeak his own disoriented self and sense of displacement, even when depicting Medora, which pinpoints her own estrangement as well: "Her face did not fit, seemed not of this place at all" (21). Indeed, if her face "did not fit," she also claims she "was brought here when [she] was a child, and that makes [her] not from here" (24).

On the other hand, the Alaskan setting also hinges upon a mythical space which is often alluded to as "the last frontier." In her book entitled *Nature's State: Imagining Alaska as the Last Frontier*, Susan Kollin explores the popular imagery of the state by looking into the connections between the imaginary and the historical. She notes that "Alaska has been situated as a sublime wilderness area in the nation's spatial imagination. Widely regarded as the Last Frontier, Alaska is positioned to encode the nation's future, serving to reopen the western American frontier that Frederick Jackson Turner declared closed in the 1890s" (Kollin 5) In *Frontier: American Literature and the American West*, Edwin Fussell purports that the frontier, contrary to its political realities, relied on the literary exploration of an in-between space, and was "the threshold of creation-destruction" (Fussell 81) or even a neutral ground where "reconciling opposites" becomes possible. (Fussell 17)

By setting the plot of *Hold the Dark* in Alaska, William Giraldi may have attempted to conjure up the mythopoetic and cultural images associated with the state. But I would contend that the novel first and foremost attempts to sketch out an imaginary in-between space, in which the author can give free rein to his "audacity," which would be in line with his exploring the monstrous:

> No honest, self-aware novelist denies the unconscious mechanisms at work and play in his composition, or attempts to refute the insistent facts of his past, facts that heave and breathe in his sentences, in his apprehensions of art and life. Like my Catholic youth, both of my novels are flesh-obsessed, preoccupied with mythos and monsters, sin and redemption, affliction and deliverance. (Giraldi 2018: 23)

In his novels, William Giraldi's in-between, or what he calls "Creative Destruction," which can be reminiscent of Edwin Fussell's own phrasing.

The multiple linguistic and semantic conflations between *wolf* and *man/woman* partake in the blurring of frontiers between *animality* and *humanity*. Indeed, while Core's internal focalization enables the reader to sense that Medora may also suffer from a sense of estrangement, it also hints at Medora's transformative nature: "[Her face] was the pale unmarked face of a plump teenage softball player, not a woman with a dead boy and a husband at war. Her eyes were pale too. In a certain angle of lamplight they looked the sparest sheen of maize, almost gold" (*Hold the Dark* 21). The last phrase echoes the gold in the eyes of the she-wolf that Core had to reluctantly kill at Yellowstone. The novel forces the reader to adjust his or her ability to see and interpret, favoring an oblique perspective on things, as is suggested by constant shape-shifting, via the reference to a totem pole bearing "multicolored faces of bears, of wolves, of humanoid creatures he could not name [...]" (20).

The novel is fraught with tropes that interweave wolves with humans and shift the attention from one to the other to deconstruct the preconceived negative connotations that depict the wolf figure. As Core's characterization suggests, men cannot do without wolves, their existence being interdependent, as the subject reversal shows in the following sentence: "He'd begun as a nature writer, and in search of a project went north where the gray wolves found him one fall, watched him for a week as he camped and fished there" (12). Starting as the grammatical subject of the sentence, the pack of wolves then becomes the one studying and observing Core. The sentence thus plays on a reversal of roles in the usual wildlife observation scene, which results in Core being gazed at by the pack of wolves. The writing style thus encourages the reader to shift his/her attention from wolves to men in finding the culprit. It also helps question our own ways of categorizing and delineating reality and what it means to be human or animal.

The novel thus induces the reader to reconsider what he/she would consider as "other." Imbued with a Gothic heritage which hinges upon an exploration of otherness through the uncanny, *Hold the Dark* also strives to avoid a close portrait of the human consciousness but, instead, delves into a de-territorializing perspective on writing and on living beings. Drawing from Deleuzian essays on space, literature partakes in a de-territorializing process which, if embedded in spatial vernacular, is also meant to question dichotomies, and attempts to outline a form of becoming-animal. In their book *Deleuze and the Animal*, Patricia MacCormack and Colin Gardner assert that:

> Unlike Spinoza, Deleuze includes the human in the category of animal (as a sketch of 'life'), and claims that the *Ethics* is an *ethology*' (1988: 27). Deleuze's critique of the abstract semblances Man makes upon the world as a teeming series of relations of expression and affects shows anthropocentric discourse as both unethical and a denial of the affirmation of life. Denying species continues this point, where Deleuze shows that within each life each organism's own affects and expressions already teem with each other, so his reading of Spinozan ethics emphasises the ecosophical nature of what philosophies of ethics take into account – all relations, all affects, all expressions and ways in which to increase rather than diminish each life as a haecceity free to express (4).

In the novel, the linguistic play on transgression allows for multiple crossings between humans and wolves and shatters the idea of their being at odds with one another. It also opens multiple transfigurative possibilities wherein man is depicted as an animal, both the human and the animal being made of affects and constantly de-territorializing each other.

The wolf-mask serves as an example of interspecies hybridity and is used by both Medora and Vernon as a prop which allows them to paradoxically act and unleash their true nature, even if that means paradoxically hiding behind a mask.

The masked face creates a utopian space, a space of otherness of which the body becomes a part, making it a fragment of imaginary space. By using the mask given to her by a wanderer, Medora both embraces this otherness that makes her and her brother stand out, but also hides beneath it when – we can only assume – she committed the unredeemable act, infanticide. When she sets sail into the Alaskan wilderness, the only clue she leaves behind is the mask. Thanks to it, Vernon understands what she has done, as if the wolf face reveals some inner secret about the common animality that drives them both and which they use to shy away from their trauma and responsibility.

In addition to the Slones, many secondary characters who hold a kind of prophetic power also seem to navigate between a human shape and an animal one, in a constant becoming-other process which is fostered by language. This process can be envisioned thanks to the garments worn by the Yupik hag and the shaman mainly, who seem part human part animal. Their polymorphic figures are reminiscent of the Yu'pik folklore, mentioned several times in the novel, whose tales are peppered with shapeshifting creatures: "He saw the red-orange radiance on her jowls, her creature garment thick and soiled, pungent-looking in the firelight, an anorak a century old. Her feet and shins were sheathed in moose-hide mukluks" (39). Core's depiction of the hag can be contrasted with that of the shaman, who is described as a possible impostor through Vernon's eyes: "Donned wholly in gray wolf pelt, with white man's skin and untrimmed hair still dark despite his age, he seemed a make-believe shaman. The wolf's tail was still attached to his guise, its fanged head pulled low over his own for a hood" (104). Being disguised contributes to their bodies' displacement, as they create a new space in which they can live their own fantasies and be immersed in other utopian rituals.

While the main characters strive to regain some sense of place, the uncanny permeates their reality and offsets what they think they know. By staging the sense of otherness, the moonscape reflects the constant becoming process and shapeshifting of both humans and wolves, whose definition becomes blurred and sheds light on their interdependence. The novel thus enables hybrid living beings to emerge, showcasing the complexity of defining humanity without considering other living forms. Blending humans and wolves constantly, it questions human nature without setting it apart. By decentering it and taking into consideration mythopoetic folklore, the novel hinge upon and revolves around more permeable ontological and epistemological definitions.

Conclusions

The following quote encapsulates the role of language and literature in the blurring of boundaries in this contemporary Gothic novel: "You would bar the door against the wolf, why not more against beasts with the souls of damned men, against men who would damn themselves to beasts?" (*Hold the Dark* 40) Suggesting man's downfall in this god-forsaken place, the hag conveys her own mistrust towards men's nature by relying on her own Yu'pik culture, but also imbuing her words with religious undertones she teased out from the books brought by the Christians. Besides, she points out the irony of the Christians' arrival, who not only brought books, but also the plague, hinting partly at the source of the place's ruin, which is rooted in human action and intervention: "They came with books and they came with the plague" (40). The chiasmus of the quote once more highlights the becoming-other process at stake in the novel, pointing towards an empathetic apprehension of the animal, blurring its definitions. As such, the novel does not pretend it can comprehensively understand the wolf, despite the wolf-expert's internal focalization, and the animal remains "the unknowable other" (MacCormack and Gardner 4), but so is the human consciousness. Instead, the textual space of the novel itself offers a threshold to a new world, which is made of linguistic and narrative possibilities, all ingrained in a process of continuous de-territorialization:

> To become animal is to participate in movement, to stake out a path of escape in all its positivity, to cross a threshold, to reach a continuum of intensities that are valuable only in themselves, to find a world of pure intensities where all forms come undone, as do all the significations, signifiers, and signifieds, to the benefit of an unformed matter of deterritorialized flux, of nonsignifying signs. (Deleuze and Guattari 13)

The novel points out the only thing it can do, which is, through the lens of Core's partial representation, to open up possibilities of being and becoming, which foster a sense of movement and to rely on affects and intensities through a de-territorialized language – playing on signifying – which contributes to a rhizomatic view of the environment and its inhabitants.

References

Buell, Lawrence. *The Environmental Imagination. Thoreau, Nature Writing, and the Formation of American Culture*. Cambridge, MA: Harvard UP, 1995. Print.

Deleuze, Gilles. *Cinema 1: The Movement-Image*, trans. H. Tomlinson and B. Habberjam, Minneapolis: University of Minnesota Press, 1986. Print.

Fussell, Edwin. [1965] *Frontier: American Literature and the American West.* Princeton: Princeton University Press. 2016. Web.

Giraldi, William. *American Audacity: In Defense of Literary Daring.* New York: Liverlight Publishing Corporation, 2018. Print.

Giraldi, William. *Hold the Dark.* New York: Liverlight Publishing Corporation, 2014. Print.

Kollin, Susan. *Nature's State: Imagining Alaska as the Last Frontier.* Chapel Hill & London: University of North Carolina Press, 2001. Print.

MacCormack, Patricia, and Colin Gardner. "Introduction." *Deleuze and the Animal.* Eds. Patricia MacCormack and Colin Gardner. Edinburgh University Press, 2017, pp. 1–24. Print.

Royle, Nicholas. *The Uncanny.* Manchester: Manchester University Press, 2003. Print.

Slotkin, Richard. *Regeneration Through Violence: The Mythology of the American Frontier, 1600–1860.* New York: Harper Perennial, 1996. Print.

Spooner, Catherine. *Contemporary Gothic.* London: Reaktion Books, 2006. Print.

Carla Francellini

Ecological Consciousness and Environmental Migration through Animal and Machine Tropes in John Steinbeck's *The Grapes of Wrath*

Abstract: This essay examines Steinbeck's animal and machine imagery in *The Grapes of Wrath* to describe the Joads' devastating journey from Oklahoma to California through the Mojave Desert. The novel takes a two-pronged approach to exploring the causes of environmental destruction and the Dust Bowl migration, touching on economic gain and class conflict issues. While questioning the impact of mechanized farming, Steinbeck criticizes the tenants' excessive plowing, which contributed to soil damage.

Introduction

In his novel *The Grapes of Wrath* (1939), John Steinbeck explores the link between economic crisis and climatic displacement, depicting environmental migration as a result of both, while emphasizing how human actions and technology can damage the environment and the land on a smaller scale. The writer's use of animal and machine symbolism is crucial to the novel's structure and meaning, as it connects the Joads' journey with the intercalary chapters that give the story a universal significance. Animals and machines, therefore, play an essential role in conveying Steinbeck's most pregnant symbolic and allegorical meanings, as is the case with the land turtle (Steinbeck 16-18), a clear example of "mediating material object for tying together Tom, Casy, and the plot, a kind of externalizing vessel, or 'symbol'" (Burke 68–69).

Throughout U.S. and world history, natural disasters have been known to cause social crises and environmental migration. Zora Neale Hurston's novel, *Their Eyes Were Watching God* (1937), expertly depicts the tragic effects of the 1928 Okeechobee Hurricane on migrant laborers in the Southern United States. Similarly, Steinbeck's masterpiece highlights the environmental displacement resulting from dust storms that forced thousands of sharecroppers from Oklahoma to California. These novels provide insight into the causes of such phenomena while "showing the effects of ideological and governmental forces on seemingly natural disasters" (Manzella 67). Additionally, extreme weather events linked to climate change, like droughts and floods, serve to "remind us of how citizenship rights are denied at various points in U.S. history as well as how the category of *internally displaced person* needs to be rethought" (Manzella 67).

On April 14, 1935, a date that would pass to history as Black Sunday, a massive dust storm, described as "a raggedly-topped formation on the move" (Egan 7), covered the horizon and engulfed the plains of the United States.

> The air crackled with electricity. Snap. Snap. Snap. Birds screeched and dashed for cover. As the black wall approached, car radios clicked off, overwhelmed by the static. Ignitions shorted out. Waves of sand, like ocean water rising over a ship's prow, swept over the roads. Cars went into ditches. A train derailed. (Egan 7)

This was the first of many storms that covered the North-to-South stretch of land from North Dakota to Texas and only a few days later, it deposited dust onto the East Coast and into the Atlantic Ocean. The expression *Dust Bowl* was coined by Robert Geiger, an Associated Press reporter, to refer to the area, including southwestern Kansas, Oklahoma panhandle, Texas panhandle, northeastern New Mexico, and southeastern Colorado, devastated by nearly a decade of drought and soil erosion. The region of the Great Plains, once known for its rich, fertile prairie soil, which had taken thousands of years to build up, was overgrazed after the Civil War by cattlemen, soon replaced by farmers, that, by World War I, overplowed the land with wheat.

In the 1920s, thousands of additional farmers migrated to the area, and powerful gasoline tractors quickly removed the remaining native prairie grasses (Manzella 69). An eight-year drought started in 1931, and winter winds took their toll on the terrain, unprotected by indigenous grasses that once grew there (Hämäläein 2016). Massive dust storms – 14 in 1932, 38 in 1933, 110 in 1934 – known as "black blizzards," combined with drought to ravage the area and destroy the crops. Cattle starved or were sold since there were no grasses to feed them. People wore gauze masks and put wet sheets over their windows. Outside, the dust buried cars and homes. Soon the Great Plains turned into a desert as over 100 million acres of intensively cultivated farmland lost most of its topsoil.

In 1935, President Franklin D. Roosevelt created the Drought Relief Service, which offered relief checks and food handouts, but did not help the land. Mysterious illnesses like *dust pneumonia* and *brown plague* began to surface. 2.5 million people – not only farmers but also business people, teachers, and medical professionals – left their homes by 1940, often heading West in search of a better future.[43]

---

43 For a wider analysis of environmental migrations, see Wood's "Ecomigration" (2001) and Shillinglaw's *On Reading* The Grapes of Wrath (2014).

The "Dispossessed" Joads and the Other(s)

Focusing on the brutal living conditions of the *dispossessed* Joads during their journey and dire life in California, Steinbeck's novel opens an extended reflection on the issues of *land ownership* and *civil rights*, both profoundly influenced and shaped by the Great Depression and the Dust Bowl migration. The Joads' loss of their "home" in Oklahoma, their tools, their family's older members, and their cultural identity is a powerful example of the dispossession many Americans – migrants within their own country – experienced during this time (Francellini 2023). Henry Nash Smith's myth of the "yeoman" – an independent farmer who cultivates his land, equating power and success with landownership and self-possession – was proven unrealistic during this era. As Manzella maintains, "the Dust Bowl migration exposes a specific incarnation of the 'white settler society myth,' that is the idea that white Europeans came to a 'blank slate' (in this case, what would become the United States) and 'developed' it without the aid of other people"[44] (69). The historical reality of U.S. migrant workers on the West Coast during the Great Depression was utterly different. While "[f]or generations, non-native-born workers – Mexican, Filipino, Asian – ha[d] planted, pruned, and harvested California's farmlands and orchards," from about 1935 to the early years of World War II, "dispossessed Anglo-Americans, mainly from the Great Plains, accomplished the main field labor in California's fertile valleys" (Wixson xvi). The same category of '*whiteness*' changed during this extended state of migratory emergency, while the notions of landownership concerning class and race were profoundly altered. The more consistent racial group on the move comprised Americans who settled in the Great Plains during the XIX century. This conjuncture brought to the national consciousness the plight of *migrant workers* and exposed the ideological contradictions regarding the lack of care for internally *displaced people*.

Even though we will not deal with the many social and historical issues connected with the whole question, we still mean to highlight how the notions of *landownership, citizenship,* and *power* in *The Grapes of Wrath* called upon concepts of governmentality, bare life, and precarity, as discussed by Giorgio Agamben, Judith Butler, and Alexander Weheliye. If in Hurston's novel the notions of landownership and rights were shaped by the foundation of slavery that led to

---

44 The quotes in Manzella's text are from Razack, *Race, Space, and the Law* (2002), p. 3, and Chris Vials, *Realism for the Masses* (2013), pp. 110–148. See also Carlos Bulosan's *America Is in the Heart* (1946).

limited citizenship for African Americans, in Steinbeck's masterpiece the same issues are refashioned by tracing the nineteenth-century settlement of the Great Plains and the subsequent disaster of the Dust Bowl migration. Consequently, in its Depression-era manifestation, the mythic yeoman no longer opposes the plantation slave system, as in the nineteenth century, but the capitalist incorporation of farming on the West Coast.

Nature, man, technology

In *The Grapes of Wrath*, Steinbeck delves into the critical connection between Nature and technology, emphasizing man's paramount responsibility towards the environment and ingeniously introducing this crucial topic long before the advent of ecocriticism. Through the two dominant motifs of *machines* and *animals* – deeply interconnected with other secondary themes – Steinbeck provides a series of significant symbols and tropes that become indispensable components of the novel to elucidate the writer's unwavering stance towards the environment.

Having noticed his preoccupation with animal imagery and symbols, some critics labeled his view of man as "biological."[45] In *The Grapes of Wrath*, however, the animal motifs are used to illustrate certain aspects of human nature, but they do not suggest that humans *are* or *ought to be* like the animals he includes in his writing.[46] Even though the Okies are described as crawling across the country "like ants" (244, 249, 297), being driven "like pigs" (401, 438), and fighting "like [...] cats" (22), Steinbeck's firm belief is that they have been *forced* into a *bestial* existence by the capitalist system. Moreover, any reference to the exploited people through animalistic images is not unfavorable since the few derogatory animal tropes apply mainly to the exploiters (banks, land companies, and profiteers). Therefore, the Okies' animalism is not presented as their fault, but due to the machine economy's encroachments.

Some animal tropes refer to the Okies' harmless playfulness or swagger – Winfield Joad is "kid-wild and calfish" (99), and Al acts like "a dung-hill rooster" (441) – but in most cases, they are used to characterize the Okies' social plight and desire for justice. "[P]iled in John's house like gophers in a winter burrow"

---

45 For a harsh critique of Steinbeck's "biological view," see Wilson, "The Boys in the Back Room [5. John Steinbeck]" (1956), Lisca, *The Wide World of John Steinbeck* (1958), and Bracher, "Steinbeck and the Biological View of Man" (1957).
46 The Okies' employers treat their farm animals better than they do with the Okies: "Fella had a team of horses, had to use 'em to plow an' cultivate an' mow, wouldn' think a turnin' 'em out to starve when they wasn't workin'. Them's horses – we're men." (454)

(49), "lift[ing] their heavy feet like draft horses" (462), the Joads are described as "beating their wings like a bird in an attic [...] tryin' ta get out" (261) as they desperately attempt to search a better future if not claim their right to the *pursuit of happiness*, as outlined in the American Declaration of Independence. Steinbeck also means to underscore some critical issues in his narrative by resorting to animal images, such as the primal drives of human sex or self-preservation. Muley Graves, for example, describes himself during his first sexual act as "stampin' an' jerkin' an' snortin'' like a buck deer, randy as a billygoat" (53) and a young, virile Al Joad is said to have been "a-billygoatin' aroun' the country" (86). As the Joads depart from their Oklahoma land, their pets are left behind and slowly revert into a primitive state, resembling their wild ancestors. This natural reversion hints at the harsh living conditions that the Okies face due to the challenges and hostility they encounter.

> When the folks first left, and the evening of the first day came, the hunting cats slouched in from the fields and mewed on the porch. And when no one came out, the cats crept through the open doors and walked mewing through the empty rooms. And then they went back to the fields and were wild cats from then on, hunting gophers and field mice, and sleeping in ditches in the daytime. [...] The wild cats crept in from the fields at night, but they did not mew at the doorstep any more. They moved like shadows of a cloud across the moon, into the rooms to hunt the mice.[47] (121–122)

The cats that Tom and Casy found on the Joads' abandoned farm foreshadowed the appearance of Muley, who had reverted to a kind of wild state of Nature. Even "[t]he two rangy shepherd dogs" in Uncle John's place – that "trotted up pleasantly until they caught the scent of strangers" and then "backed cautiously away, watchful, their tails moving slowly and tentatively in the air, but their eyes and noses quick for animosity or danger" (76) – acted like people who were cautious when faced with strangers or new situations. These animals serve as symbols mirroring the sharecroppers' struggles in the fast-paced, mechanized industrial economy, as the scene of the "jackrabbit [...] caught in the lights" proves.

> [H]e bounced along ahead, cruising easily, his great ears flopping with every jump. Now and then he tried to break off the road, but the wall of darkness thrust him back. Far ahead bright headlights appeared and bore down on them. The rabbit hesitated, faltered, then turned and bolted toward the lesser lights of the Dodge. There was a small soft jolt as he went under the wheels. The oncoming car swished by. (193)

---

47 The Joads' love for pets symbolizes human sympathy for life and Nature. When their fortunes are at their lowest ebb, Ma still holds hopes for a pleasant future: "'Wisht we had a dog,' Ruthie said. [Ma replied] 'We'll have a dog; have a cat too.'" (458)

The animals in Steinbeck's novel and their associated colors are often used as omens and prophetic signs, as in the highly dramatic passage before the Okies leave for California, where "[t]he shadow of a buzzard slid across the earth, and the family all looked up at the sailing black bird" (174). Animal tropes can also denote violence or depravity in human behavior towards other men or the environment. A tractor hitting a sharecropper's cabin "give[s] her a shake like a dog shakes a rat" (48); Muley used to be "mean like a wolf" when he was a "hunter," and then "mean like a weasel" once he "get[s] hunted" (60); and, finally, Ma Joad describes Purty Boy Floyd's career as comparable to a maddened animal at bay: "they shot at him like a varmint, an' he shot back, an' then they run him like a coyote, an' him a-snappin' an' a-snarlin', mean as a lobo" (80).

People's nature or character are also frequently represented by their reaction to the treatment of lower animals. As Tom and Casy walk along the dusty road, a gopher snake wriggles across their path: Tom, who is neither cruel nor perverse, peers at it, sees it is harmless, and, therefore, lets it go (72). Later in the novel, a "rattlesnake crawled across the road," and Tom reacts by "hit[ting] it and br[eaking] it and le[aving] it squirming" (241). The two scenes – astonishingly similar, though opposed – attest to Tom's violent temperament, but also to his ability to discern good from evil.

Throughout the novel machines and animals frequently merge, symbolizing the destructive force found in Nature (Casey 1997). As Steinbeck describes the banks as *monstrous* animals – "A bank isn't like a man" (35), they are "machines and masters all at the same time" (33) – and the tractors as "[s]nub-nosed *monsters*, raising the dust and sticking their snouts into it" (37), he's intending a harsh critique of technological aggression resorting to animalistic comparisons and metaphors. Everything becomes evident when the tractors, "moving like insects, having the incredible strength of insects," crawl menacingly across the land, "straight down the country, across the country, through fences, through dooryards, in and out of gullies in straight lines." Avoiding contact with the land – the tractors "did not run on the ground, but on their own roadbeds!" – "[t]hey ignored hills and gulches, watercourses, fences, houses" (37).

The machines' lack of contact with the land symbolizes a lack of knowledge of the land itself, a fact that triggers an antagonistic relationship with the environment on which human beings depend.[48] The assimilation between men and machines also displays Steinbeck's idea that they both can be violent and

---

48 The expressions *tractored out* and *tractored off* recur in the text as "prominent figure[s] of speech [...] to express the Okies' plight in being forced from their plots of land by the mechanical monstrosity of industrialized farming." (Griffin, Freedman, "Machines and Animals: Pervasive Motifs in *The Grapes of Wrath*." 2008, p. 570)

destructive: in the novel's final, the weary men trying to build a bank of earth to hold back the flood are described as "work[ing] jerkily, like machines" (461). Even the tractor's driver – once an old acquaintance of the farmers – appears to his fellows as a part of the machine, a realization emphasizing technology's uncanny side. !

> The man sitting in the iron seat did not look like a man; gloved, goggled, rubber dust mask over nose and mouth, he was a part of the monster, a robot in the seat. [...] the monster that build the tractor, the monster that sent the tractor out, had somehow got into the driver's hands, into his brain and muscle, had goggled him and muzzled him – goggled his mouth, muzzled his speech, goggled his perception, muzzled his protest. (37)

Sigmund Freud's notion of "*uncanny*" perfectly applies to the description of machinery and *forced* modernization in a kind of agriculture characterized by violent and raping practices.[49] Described as frightfully non-human and unfamiliar (*Unheimlich*), "the man in the iron seat" opens to the traumatic possibility that a human being might not be "human" at all, while the word "monster" insists on men's and machines' destructive power (Freud 219–222).

Many mechanical devices are presented as objects suggesting abstract qualities. This is the case of the "huge red transport truck" (7), with its newness, mobility, efficiency, and even inhumanity (the sign on its side reads: "No riders are admitted"), a clear allegory of the mechanical-industrial era Okies are bound to experience. The used-car business (64-69) – preying on the Okies' need to move out and move quickly – symbolizes the exploitation of those not involved in society's mechanization on a large scale. The Joads' truck – around which the members of the family gather for their final council before leaving Oklahoma – becomes their "most important place," the only "active thing, the living principle" once their "house [is] dead, and the fields [are] dead." (104). At the same time, it is a symbol of their precariousness and inability to move efficiently or in style towards California: "The engine was noisy, full of little clashings, and the brake rods banged. There was a wooden creaking from the wheels, and a thin jet of steam escaped through a hole in the top of the radiator cap" (103). Their truck becomes "home" to the Joads, whose identification with the machine is so deep that when their physical condition worsens, so naturally does that of the truck: "the right head light blinked on and off from a bad connection" (420). When the disastrous flood begins, the "water fouled the ignition wires and water fouled the carburetors" of the old cars "[b]eside the tents" (453). When the rains

---

49 The German word "*unheimlich*" literally means "unhomely," or "uncanny."

become heavier, "trucks and automobiles deep in the slowly moving water," as Pa and Ma "slipped into the water" (472), together with their hopes to find work in the rainy season.

Climate changes and circular narrative patterns

Climate changes as a devastating manifestation of Nature's power are another significant issue in *The Grapes of Wrath* and provide a frame for the narrative through a circular pattern, linking the dust storm at the novel's beginning to the flood at the end. Both highly reminiscent of their biblical archetypes, the two phenomena – as described in Steinbeck's novel – insist on the idea of Nature governed by Chaos, a great force that can give life and take it. On the contrary, technology and machines are run by numbers, order, and efficiency. However, Nature's destructive power and unrecognizable patterns are more familiar to the tenants than the destruction of their lives brought about by technology, machines, and banks, all being part of the "*uncanny*." The unnatural straight lines made by the "rapist" tractors reiterate the idea of "*cutting*" as a kind of violent "surgery." At the same time, violence and sexual allusions combine in this passage to convey man's devastating impact on the land.

> Behind the tractor rolled the shining disks, cutting the earth with blades – not plowing but surgery, pushing the cut earth to the right where the second row of disks cut it and pushed it to the left; slicing blades shining, polished by the cut earth. And pulled behind the disks, the harrows combing with iron teeth so that the little clods broke up and the earth lay smooth. Behind the harrows, the long seeders – twelve curved iron penes erected in the foundry, orgasms set by gears, raping methodically, raping without passion. (38)

The drought and the flood round off the book and make it almost a *living* thing, complete and extended, very distant from the kind of narrative consisting of a straight thin line between the two arbitrary points of its beginning and end. The novel's circularity implies a careful symbolic structure, both naturalistic and experiential. In the first chapter, Steinbeck portrays how Nature's forces can transform everything into death and dust while the novel's main characters struggle to eke out a living amidst the drought. The writer's description of natural disasters innovatively creates a clear causality between climate, devastating farming practices, aggressive banks and corporation policies, and the Dust Bowl crisis. Steinbeck also emphasizes how, despite its chaotic and extreme manifestations, Nature (in the broadest meaning of the word) ends up providing a sense of "home" (*Heimlich*) to the tenants while technology, with its potential for progress and change, remains a source of uneasiness (*Unheimlich*).

The Joads' and the other sharecroppers' reaction to natural disasters is the same as the animals caught in similar situations of danger, so much so that fearing that the corn should die, they do what animals would. They look for a shelter – "[m]en and women *huddled* in their houses" (4) – and wait for the storm to be over: "lying in their beds, [the people] heard the wind stop." (5) Following a similar pattern, in the last interchapter of the book, the already shattered Joad family faces the flood. Deprived of their car, their most prized and essential possession, they fear starvation to death – "And gradually the greatest terror of all came along. They ain't gonna be no kinda work for three months" – as they spontaneously band together with other migrants to handle the disaster: "And when the puddles formed, the men went out in the rain with shovels and built little dikes around the tents. The beating rain worked at the canvas until it penetrated and sent streams down" (453).

As they had waited for the dust storm to be over, they waited for the rain to stop falling while sheltering in their tents: "the migrant people huddled in their tents, saying, It'll soon be over, and asking, How long's it likely to go on?" (452–453). The sharecroppers act in the same script, even though they face two opposite extreme natural disasters, prompted by their instinct of self-preservation. Their reaction proves that the Joads persist in a condition of animality, into which they are bound to merge more and more as they live and proliferate in a capitalist system.

The land turtle

The animal that best symbolizes the Joads' resilience and primal drive for life – a crucial issue in the novel – is the land turtle that Tom finds on his way home at the beginning of the book[50]. With his primitive and constant movement, this reptile foreshadows the Okies' perseverance in reaching California, no matter how arduous the trek is. Tom's slow progress down the dirt road is symbolized by a reptile that, despite facing various obstacles – being trapped in Tom's coat for a while, a motorist trying to hit him, a cat attacking him– eventually succeeds in crossing the highway. Tom's description – he "plodded along, dragging his cloud of dust behind him" and "moved ahead […], dragging his heels a little in the dust" – recalls the land turtle's movements: "the high-doomed shell of a land turtle crawl[ed] slowly along the dust, its legs working stiffly and jerkingly" (20).

---

50 The tortoise is a powerful symbol of the contrast between good and evil in Melville's *Las Encantadas, or the Enchanted Isles* (1854).

A kind of similarity – based on the notions of freedom and independence – also links the land turtle and the controversial character of Casy, as the ex-preacher correctly observes.

> Nobody can't keep a turtle though. They work at it and work at it, and at last one day they get out and away they go – off somewheres. It's like me. I wouldn' take the good ol' gospel that was just layin' there to my hand. I got to be pickin' at it an' workin' at it until I got it all tore down. (23)

However, unlike the turtle, Casy appears unsure of the path he wants to take in life as he openly confesses: "I have the call to lead the people an' no place to lead 'em'"; "'That's right, he's [the turtle's] goin' someplace. Me – I don't know where I'm goin'" (23–24). The land turtle's continuous movement alludes to how all life naturally seeks to go somewhere through an instinctive urge for self-realization. The scene in the third chapter, with the turtle carrying "one head of wild oats clamped into the shell by a front leg," is a crucial introduction to the theme of environmental interdependence. The reptile confidently marches towards the highway, skillfully gripping the wild oat head by its stem with its front legs. It then drops the head, causing "the wild oat head f[a]ll out and three of the spearhead seeds stuck in the ground," while the "shell drag[s] dirt over the[m]" (17–18). This powerful symbolism effectively emphasizes the interconnection between humans and Nature, a recurring theme throughout the novel.

The scenes of the tractor and its driver, that of Muley Graves with his tales of lonely scavenging on the deserted land, the second-hand car dealer, Highway 66, the Joads' truck, the empty, abandoned houses, the federal camp, the Hooverville camp, Noah's departure, the death and burial of Granpa, Casy's death, and the flood can all be read as symbols of the Okies' basic attitudes, conflicts, and primal drives in life. Some of these attitudes are more social than others, apparently more universal or epic. The first ones range from the positive values of folk fellowship and morality (the new Law of the Road) to group action and democracy-in-process. The last ones are more significant to our research since, at the level of the *universally human*, Steinbeck highlights once again man's dependence on the primal elements (water, sun, fire, land) and emphasizes the epic nature of sex, womanhood, family life, death, and mutualism of spirit. The novel finally suggests that Tom and Rosasharn identify with humanity. Underscoring the epic idea that all men are brothers since all men belong to the Race of Man, Steinbeck goes beyond any biological approach to ethics, emphasizing the spirit's transcendent yet absolute unity. Through the skillful interweaving of the interchapters with the narrative chapters, Steinbeck's search for spiritual values looks inside human experience, Nature, and life processes, being teleological only in the word's scientific (not metaphysical) meaning of the term. His naturalism goes beyond mechanistic determinism and lifts the biology of stimulus-response to the biology of spirit (Carlson 175). Steinbeck's idea of epic mutualism is not romantic, mystic, or Christian and seems more an experiential discovery of the process by which "physiological man" becomes the "whole man" (*The Sea of Cortez* 87) through the "humanistic integration of the knowledge of man, made available by modern science, philosophy, art" (Carlson 175).

The "evil" tractors

Frequently depicted as evil objects that "throw men out of work," the tractors also convey the feeling (and the fact) that farming is being transformed into a mechanized industry – "[t]he great owners formed associations for protection, and they met to discuss ways to intimidate, to kill, to gas" (249). The process is irreversible – "Heard 'bout the new cotton-pickin' machine?" (426) – and nothing can stop it, as the episode of Granpa shooting at a tractor that keeps moving across the Joads' land proves (48). Steinbeck does not convey the feeling that machines are *automatically* or *necessarily* evil since they are just *instruments that, in* the hands of the right people, could become instruments of good fortune. In the same way, the tractors are not inherently evil but just symptoms of unfair exploitation. A clear metaphor for this way of conceiving machines is offered in the scene of the turtle crossing the highway, with one driver swinging to avoid him and another swerving to hit him (17). Analogously, the tractors are not inherently evil but just symptoms of unfair exploitation.

> Is a tractor bad? Is the power that turns the long furrows wrong? If this tractor were ours it would be good – not mine, but ours. If our tractor turned the long furrows of our land, it would be good. Not my land, but ours. We could love that tractor then as we have loved this land when it was ours. (157)

Pieces of machinery, like science and technology, can enhance agriculture and develop better crops.[51] However, they are only enough for progress with human understanding and cooperation, and people need help to come to terms with the "*machine age" in which* they live. Al Joad's relationship with the truck symbolizes the complexities of getting accommodated in a new era. Since he knows

---

51 "Behind the fruitfulness are men of understanding and knowledge and skill, men who experiment with seed, endlessly developing the techniques for greater crops of plants whose roots will resist the million enemies of the earth: the molds, the insects, the rusts, the blights." (362)

about motors and takes care of the truck – "Al was one with [the truck's] engine, every nerve listening for weaknesses, for the thumps or squeals, hums and chattering [...] He had become the soul of the car" (128) –, he is admitted to a place of responsibility in the family.[52] Toward the end of the book, Ma's request for Al to stay with the family emphasizes the importance of the truck in the Joads' lives. Moreover, even though the flood puts the truck out of action, the novel ends on a hopeful note of human sharing that foreshadows the possibility that the Okies' children will eventually assimilate into a machine-oriented society.

The Grapes of Wrath suggests a twofold approach to the circumstances leading to environmental displacement and the Dust Bowl migration. On one side, it tackles the issues of economic profit and class struggle, opening the question of mechanized agriculture and farming. Conversely, the novel points to the tenants' methodical plowing, partially blaming it for damaging *"their"* land's soil. One of the outstanding achievements of Steinbeck's masterpiece is its stand on the environment, denouncing the abuses committed by banks, corporations, and tenant farmers. The tractors – Steinbeck seems to suggest – are not the only villains in the novel since the tenants have their responsibilities. Man, though, can continue his progress like the turtle, becoming aware of his goals and deliberately employing new devices to attain them. Nevertheless, man's progress does not have to be blind, for he can couple human knowledge with human love and manipulate science and technology to make the betterment of himself and all his fellows possible within a correct and healthy relationship with the environment.

The Okies' "Counter-Pastoral"

A kind of *"Counter-Pastoral"* seems to linger in Steinbeck's narrative as the migrants claim their attachment to the land but also exhibit selfish exploitation and disregard towards it. Highlighting how generations of Oklahoman tenants and sharecroppers had plowed the earth and reaped its harvest without thought of ever replenishing and preserving it, the novel displays how the migrant's pastoral dream of harmony with Nature may even sound appealing, but it is not authentic. Looking back at their past with nostalgia, the tenants idealize their relationship with the land to the point of self-deception, as they

---

52 Casy tells Tom: "Funny how you fellas can fix a car. Jus' light right in an' fix her. I couldn't fix no car, not even now when I seen you do it." And Tom replies: "Got to grow into her when you're a little kid, [...] It ain't jus' knowin'. It's more'n that. Kids now can tear down a car 'thout even thinkin' about it." (193)

fail to remember the hardships they faced while living and working on it. Ironically, although technology corrupted the Okies' "American Pastoral"[53], their dream was already tainted by a history of destructive practices towards Nature, ending up being abusive in their long-term effects. Some characters end up acknowledging the harm produced to the land by the damaging crops. Muley, for example, notes: "Never was much good 'cept for grazing. Never should a broke her up. An' now she's cottoned damn near to death" (50). Still, no action can follow such an enlightening realization.

While denouncing both technology and the profit system behind it as uncanny and devastating, Steinbeck also condemns disregarding human attitudes towards Nature. Had the Okies remained in their land for the rest of their lives, they would have continued to abuse the soil, as they had been doing for decades before the dust storms. Therefore, their idea of blissful happiness and closeness with the land is based on its savage exploitation in Oklahoma – where the land is perceived as *"theirs"* since their ancestors fought to protect it – and California.

The irony in the tenants' *pastoral* becomes evident in the attitude of the Joads' children towards animals. When the only dog the Joads take with them on their trip is hit by a car on the highway, Ruthie and Winfield are playing in the cornfield by the gas station with a handful of reptile eggs. As soon as they realize what has happened to *"their"* dog, Ruthie loses curiosity about the eggs and throws them away, effectively destroying them. Her contradictory conduct symbolizes men's conflicting actions and emotions: though sympathetic toward what they perceive as *"theirs" (the* dog), they remorselessly destroy all the rest (the reptile eggs). The tenants are guilty of the same sins committed by the tractor drivers and the owners, relentlessly plowing the earth until it is broken and barren.

Conclusion

Since Steinbeck approached more openly the issue of individual responsibility towards the environment and the interconnectedness of man and Nature in his later works – *The Log from the Sea of Cortez (1951), Sweet Thursday (1954),* and *Travels with Charley; in Search of America (1962), America and Americans* (1966) – some critics did not see in *The Grapes of*

---

53 For a powerful depiction of a "Counter-pastoral", see Philip Roth, *American Pastoral* (1997).

Wrath a novel ecologically oriented, nor "an environmental story" (Hedgpeth 296). Some other scholars, though, detect in the novel's structure – and especially in the interchapters –, the first germs of an ecological critique due to blossom in Steinbeck's later fiction. Depicting the Joads' odyssey, though, he had already developed an ethic about the land, a sense of how and why one should live in harmony with the land (Timmerman, 1997, Heavilin, 2017). Steinbeck's sharp critique of environmental abuse in his masterpiece is even more relevant in the 1940s since it aims at remarking on the consequences of ignoring harmful human actions on the environment. Through shocking images – the tractors *raping* the land – and metaphors applied to mechanical equipment, he emphasizes the humans' *animalism* and *wild side*.[54] *In so doing, Steinbeck* infuses in the novel a general sense of tragedy and disaster, reiterated by such secondary motifs as the "*blood*" and "*cut*" tropes – "the sun was as red as ripe new blood" (5), "the earth was bloody in [the sun's] setting light" (99), "the sun cut into the shade of the truck as noon approached" (9), "the road was cut with furrows where dust had slid and settled back into the wheel trucks" (19).

The ecological disaster was only alleviated after the rains returned in 1939 and soil conservation efforts had begun. With the rain and the new irrigation systems built to resist drought, the land grew golden with wheat again (Smith 2007). President Roosevelt signed the Soil Conservation Act on April 27, 1935 while the remaining Great Plains farmers tried new methods, and two hundred million wind-breaking trees were planted across the Great Plains to protect the land from erosion. The extensive re-plowing of the land into furrows and crop rotation resulted in a consistent reduction in the amount of soil blowing away by 1938. However, despite efforts to mitigate the effects of the drought and alluvions, they persisted and posed ongoing challenges for the community.

References

Agamben, Giorgio. *Homo Sacer: Sovereign Power and Bare Life*. Palo Alto, CA: Stanford UP, 1998. Print.

---

54 See Renata Lucena Dalmaso's article "'Modern Monsters,' Old Habits: Nature, Humans, and Technology in John Steinbeck's *The Grapes of Wrath*" (2015) for a detailed description of technology as possessing monstrous qualities.

Bracher, Frederick. "Steinbeck and the Biological View of Man." *Pacific Spectator II* (Winter 1948): 14–29. Reprinted in *Steinbeck and His Critics: A Record of 25 Years*. Eds. E.W. Tedlock, Jr., C.V. Wicker, Albuquerque: New Mexico UP, 1957. 183–196. Web.

Bulosan, Carlos. *America Is in the Heart*. Seattle: Washington UP, 1973. Print.

Burke, Kenneth. *The Philosophy of Literary Form*. New York: Random House, 1957. Print.

Butler, Judith. "Performativity, Precarity and Sexual Politics." *AIBR: Revista de Antropología Iberoamericana* 4.3 (2009): i–xiii. Web.

Carlson, Eric W. "Symbolism in the Grapes of Wrath." *College English* 19. 4 (1958): 172–175. Web.

Casey, Roger N. *Textual Vehicles: The Automobile in American Literature*. New York: Garland Publishing, 1997. Print.

Egan, Timothy. *The Worst Hard Time. The Untold Story of Those Who Survived The Great American Dust Bowl*. Boston: Houghton Mifflin, 2006. Print.

Francellini, Carla. "Symbols of Dispossession in John Steinbeck's *The Grapes of Wrath*." *Precarity in Culture. Precarious Lives, Uncertain Future*. Eds. Elisabetta Marino, Bootheina Majoul. Newcastle upon Tyne: Cambridge Scholars Publishing, 2023. 294–305. Print.

Freud, Sigmund. "Uncanny." (1919) *The Standard Edition of the Complete Psychological Works of Sigmund Freud*. Ed. James Strachey, vol. xvii (1917–1919). London: The Hogarth Press and the Institue of Psycho-Analysis, 1953. 219–253. Print.

Geiger, Robert. "If It Rains. . ." *Washington Evening Star* (April 15, 1935): A-2. May 9, 2023. Web.

Griffin, Robert. J., and William A. Freedman. "Machines and Animals: Pervasive Motifs in *The Grapes of Wrath*." *Journal of English and Germanic Philology* (July 1963): 569–80. Rpt. as "Machines of Industrialization Are Impersonal Monsters." *Industrialism in John Steinbeck's* The Grapes of Wrath. Louise Hawker (ed.). New York: Greenhaven Press, 2008. 81–89. Print.

Hämäläein, Pekka. "Reconstructing the Great Plains: the Long Struggle for Sovereignty and Dominance in the Heart of the Continent." Journal of the Civil War Era 6.4 (2016): 481–509. Web.

Heavilin, Barbara A. "A Sacred Bond Broken: The People versus the Land in *The Grapes of Wrath*." *The Steinbeck Review* 14.1 (2017): 23–38. Web.

Hedgpeth, Joel. "John Steinbeck: Late-Blooming Environmentalist." *Steinbeck and the Environment: Interdisciplinary Approaches*. Eds. Susan F. Beegel, Susan Shillinglaw, Wesley N. Tiffney, Tuscaloosa: Alabama UP, 1997. 293–309. Print.

Hurston, Zora Neale. *Their Eyes Were Watching God*. (1937) New York: Harper Collins, 2006. Print.

Lisca, Peter. *The Wide World of John Steinbeck*. Brunswick, New Jersey: Rutgers UP, 1958. Print.

Dalmaso, Renata Lucena. "'Modern Monsters,' Old Habits: Nature, Humans, and Technology in John Steinbeck's *The Grapes of Wrath*." *The Steinbeck Review* 12 (2015): 26–38. Web.

Manzella, Abigail G.H. "The Environmental Displacement of the Dust Bowl: From the Yeoman Myth to Collective Respect and Babb's Whose Names Are Unknown." *Migrating Fictions: Gender, Race, and Citizenship in U.S. Internal Displacements*. Athens, OH: Ohio State UP, 2018: 67–108. Web.

Melville, Herman. (1854) *The Encantadas, or Enchanted Isles. The Piazza Tales*. Evanston: Illinois: Northwestern UP, 1996: 125–173. Print.

Roth, Philip. *American Pastoral*. Boston: Houghton Mifflin Harcourt, 1997.

Razack, Sherene H. (ed.) *Race, Space, and the Law: Unmapping a White Settler Society*. Toronto: Between the Lines, 2002. Print.

Shillinglaw, Susan. *On Reading* The Grapes of Wrath. New York: Penguin, 2014. Print.

Smith, Henry Nash. 1999. *Virgin Land: The American West as Symbol and Myth*. Cambridge, MA: Harvard UP, 1999. Print.

Smith, Rebecca. "Saving the Dust Bowl: 'Big Hugh' Bennett's Triumph over Tragedy." *The History Teacher* 41.1 (2007): 65–95. Web.

Steinbeck, John. *The Grapes of Wrath*. 1939. New York: Penguin, 2000. Print.

———. *Sweet Thursday (1954)*. New York: Penguin, 2008. Print.

———. *Travels with Charley; in Search of America (1962)*. New York: Penguin, 2002. Print.

———, Edward F. Ricketts. *The Log from the Sea of Cortez*. 1951. New York: Penguin, 1995. Print.

Timmerman, John. "Steinbeck's Environmental Ethic: Humanity in Harmony with the Land." *Steinbeck and the Environment: Interdisciplinary Approaches*. Eds. Susan F. Beegel, Susan Shillinglaw, and Wesley N. Tiffney, Jr. Tuscaloosa: Alabama UP, 1997: 310–322. Print.

Vials, Chris. *Realism for the Masses: Aesthetics, Popular Front Pluralism, and U.S. Culture, 1935–1947*. Jackson: Mississippi UP, 2013. Print.

Weheliye, Alexander G. *Habeas Viscus: Racializing Assemblages, Biopolitics, and Black Feminist Theories of the Human.* Durham, NC: Duke UP, 2014. Print.

Wilson, Edmund. "The Boys in the Back Room [5. John Steinbeck]." *A Literary Chronicle: 1Q20-1Q50* (1956): 230–239. Print.

Wixson, Douglas. "Introduction." In Sanora Babb and Dorothy Babb. *On the Dirty Plate Trail: Remembering the Dust Bowl Refugee Camps.* Ed. D. Wixson. Austin: Texas UP, 2007: 1–10. Print.

Wood, William B. "Ecomigration: Linkages between environmental change and migration." *Global Migrants, Global Refugees. Problems and Solutions.* Eds. Aristide R. Zolberg and Peter M. Benda. New York: Berghahn, 2001: 42–61. Web.

Olga Thierbach-McLean

Uncanceling the Future: Cyberpunk, Solarpunk, and the Rebellion of Eco-Optimism

Abstract: In the face of present-day environmental challenges, it often seems like we have inherited a present without a future. This eco-melancholic mood is also a product of decades-long cultural autosuggestion. Especially the dystopian forecast of *cyberpunk* fiction has conditioned audiences to think of humanity's prospects as inescapably bleak. The genre of *solarpunk* is emerging in counter-reaction to this trend. Refusing to give up on the future, it strategically cultivates optimism as a way of stimulating radical ecological reform.

Introduction

There is a growing collective sense that we are running out of future. As people across the globe are increasingly confronted with the harbingers of climate change and environmental collapse, the urgency of the ecological crisis is becoming ever more apparent. The escalating physical destruction of ecosystems has been accompanied by traumas to our collective emotional landscape, with new psychosocial phenomena such as eco melancholia and climate anxiety on the rise particularly among young people. According to a recent international survey, a majority of respondents aged 16–25 "perceive that they have no future, that humanity is doomed, and that governments are failing to respond adequately" (Hickman et al.).[55] The 2021 United Nations Climate Change Conference, which has been widely construed as a make-or-break event for environmental action, has done little to dispel such concerns. Held in yet another record year of rising temperatures and weather-related disasters, this pivotal

---

[55] Based on a total number of 10,000 children, adolescents, and young adults from ten countries, the survey found that "59% were very or extremely worried and 84% were at least moderately worried. More than 50% reported each of the following emotions: sad, anxious, angry, powerless, helpless, and guilty. More than 45% of respondents said their feelings about climate change negatively affected their daily life and functioning, and many reported a high number of negative thoughts about climate change (e.g., 75% said that they think the future is frightening and 83% said that they think people have failed to take care of the planet)."

international summit has failed to live up to public hopes for a robust political and economic course correction.

This socioeconomic inertia in the face of existential emergency corroborates the oft-quoted dictum that "it is easier to imagine the end of the world than the end of capitalism." In fact, there is growing alarm over the environmental movement itself being annexed by the economic powers that be. For instance, in his documentary *The Planet of the Humans* (2020), filmmaker Jeff Gibbs draws attention to the ways in which "green energy" has been reduced to a mere marketing label, aiding rather than counteracting the profit-fixated depletion of natural resources.[56] Along the same lines, environmental philosopher Glenn Albrecht has pointed out that the environmentalist language of renewal and resilience has been

> appropriated by forces determined to pull it into the gravitational influence of industrial society on a globalized scale. Instead of helping us rebound into configurations of successful models of living after disturbance, we are now seeing resilience being used to justify the ongoing existence of processes and activities that are driving humans to extinction. (Albrecht 304)

Habitually attributed to the inadequacies of political and economic leaders, this failure to break away from the logic of capitalism may also be the result of a limited collective imagination caused by a one-sided fictional diet. Especially when it comes to the realm of speculative fiction, critics have argued that the current prevalence of the dystopian mode has conditioned mass audiences to think of the future as inescapably bleak.[57] Among the wide range of popular near-future narratives – whether it be Cormac McCarthy's *The Road*, Margaret Atwood's *The Handmaid's Tale*, or Suzanne Collins' *The Hunger Games* – it is indeed hard to think of an example that does not forecast societal decline in tandem with some kind of human-caused environmental calamity. Even as such dystopian visions have often provided a language for political activism, the sheer ubiquity

---

[56] Several critics have pointed out that some of Gibb's claims are outdated in that they are based on decades-old technology. But that does not take away the fact that the efficiency of these technologies has been overstated at the time, nor does it refute the overall account of how corporations are attaching the green energy label to ecologically destructive practices.

[57] For a broader historic overview of the fluctuating popularity of the dystopian versus the utopian mode, see Sargent.

of post-apocalyptic narratives also seems to have mentally primed audiences for a fatalistic acceptance of the anthropogenic destruction of our planet.[58]

Probably more than any other genre of speculative fiction, cyberpunk has come under scrutiny in this context. When it emerged in the 1980s as a mostly North American phenomenon, it gave expression to contemporaneous apprehensions regarding the concurrent rise of the information age and neoliberal ideology. Early cyberpunk works such as William Gibson's genre-defining novel *Neuromancer* (1984) set the tone by describing corporate dystopias in which the capitalist marketplace has become the only source of meaning and value. Today, cyberpunk themes and aesthetics are so ubiquitous in literature, the visual arts, gaming, fashion, and music that they have come to seem synonymous with futurity itself. Especially contemporary Anglophone literature and cinema has been producing a constant stream of cyberpunk-themed works, drawing near-future scenarios in which technological advancement is obtained at the price of societal and environmental breakdown.

In a time when many of these dark warnings are increasingly resembling our reality, one is struck by how accurately cyberpunk fiction has predicted many of our present challenges – and how ineffective it has been in averting them. According to some voices, the main reason for this discrepancy lies in its penchant for re-diagnosing the problem rather than moving on to explore alternatives. Especially recent cyberpunk-themed releases like Steven Spielberg's *Ready Player One* (2018) or Robert Rodriguez' *Alita: Battle Angel* (2019) keep reproducing classic genre tropes such as corporate greed, unrestrained consumerism and environmental destruction while all but abandoning the sociocritical aspects of early cyberpunk. Designed to enthrall rather than to disturb, these movies' depiction of advertisement-littered megacities does not so much provoke critical reflection as it invites the audience to participate in the excitements of neoliberal information economy. By all appearances, the once rebellious spirit of cyberpunk has succumbed to what Mark Fisher has termed "capitalist realism," namely "the widespread sense that not only is capitalism the only viable political and economic system, but also that it is now impossible even to imagine a coherent alternative to it" (Fisher 2).

---

58 For instance, the red-and-white garb from the TV adaptation of Atwood's novel *The Handmaid's Tale* has been donned by women all over the world in protests for reproductive rights. Similarly, Collins' young-adult novel *The Hunger Games* has been widely used as a basis for teaching environmental justice in schools.

The present essay explores an artistic movement that is mobilizing in response to this impasse, namely the budding genre of *solarpunk*. Often defining itself in direct opposition to cyberpunk features – setting hopefulness against cynicism, communal cooperation against individual struggle, lush art nouveau aesthetic against techno-orientalist neon-noir –, its mission statement is to explore paths to a desirable future by formulating possible "solutions, not only warnings" (Springett web). By specifically highlighting its creative dialogue with cyberpunk, I will investigate the intellectual strategies employed by solarpunk to open new imaginative spaces beyond the capitalist paradigm – and thus contest the widespread view that "the future is cancelled."

From cyberspace to climate change: Shifting concerns in speculative fiction

In the final scene of Denis Villeneuve's film *Blade Runner 2049* (2017), the sequel to Ridley Scott's cyberpunk classic *Blade Runner* (1982), the camera comes to rest on the motionless face of protagonist K as he lies either dead or dying. Leading up to this, we have witnessed K's struggles with the forces of a hypercapitalist system built on the relentless commodification of living beings, whether it be children enslaved in work camps euphemized as orphanages, bioengineered humans like K who are denied full human status, or plants and animals exterminated in the name of industrial growth. Villeneuve's vision adds another layer of tristesse to the already gloomy source material. While the first movie – at least in its theatrical version – hints at the possibility of a better life outside the urban sprawl, *Blade Runner 2049* expands its vision to beyond the megacity to show that there is nowhere left to retreat to. Erratic weather patterns, elevated sea levels, scarcity of resources, and contaminated wastelands present a collage of all-encompassing ecological devastation. In a film infused with environmental themes, the glum ending has been interpreted as a reflection on the contemporary sense of systemic paralysis. As Matthew Flisfeder has observed,

> K's dilemma is that he knows full well about his status as replicant …, but nevertheless he continues to be compliant. Similarly, we are aware of the social, political, economic, and ecological problems that we face in our contemporary age and can avow this at a conscious level. Ideology, today, is no longer a matter of false consciousness. Instead, we know but nevertheless continue to act as if we didn't because there is apparently no alternative. (Flisfeder 144)

Although the enormousness of our collective challenges is daunting, the ecomelancholic zeitgeist has also turned into a self-fulfilling prophecy. Conservation psychologists have observed that the rampant defeatism does not only result in environmental apathy and inaction, but even fuels climate change denial and unrestrained material consumption as a hedonistic coping reflex. Tellingly, "this sense of despondency, that we're never going to get there, that it's too little too late" (Wushke in Mark and Di Battista 232) is spreading particularly among environmental activists. Needless to say, there is no guarantee that any efforts made at this point will be enough to avert ecological disaster. However, it is also true that staying on the current track renders it a certainty. With that in mind, there has been increasing emphasis in ecological humanities on fostering "a new set of ecological virtues, which include courage and radical hope against despair and hopelessness" (Kretz, 2017: 277). After all, as Lisa Kretz notes in "Hope in Environmental Philosophy,"

> [t]he future is shaped, in part, by our current attitudes, methods of framing, and attendant actions. Hope bridges the gulf between the beliefs and actions of today and possibilities for tomorrow. If positive moral action is the goal, hope is a vital concept for underwriting ecological philosophy and inspiring positive ecological action. (Kretz, 2013 web.)

In other words, to purposefully aspire to something we need to be able to imagine it first. Optimism thus serves as an epistemic strategy for unlocking new conceptual territory. Especially in view of the enormous impact of pop-cultural tropes on the framing of social discourses, "we clearly need, within popular culture, visions and memes of a terranascent future" (Albrecht 303).

Supplying such visions is the declared goal of the solarpunk movement, which understands itself as a "rebellion against the structural pessimism of how the future will be" (Owens web). The term solarpunk is commonly traced to a 2008 blog post titled "From Steampunk to Solarpunk," where the anonymous author calls for a new type of speculative fiction that actively engages with the question of how humanity could transition to a sustainable civilization (Republic of Bees). Unlike with cyberpunk, the artistic category of solarpunk thus emerged before the actual appearance of a genre-galvanizing novel or film. The movement instead started coalescing through the exchange of ideas in interactive online spaces, most notably the social networking website tumblr.com which has become a major hub of the solarpunk community. Particularly since 2012, solarpunk has been gaining traction as a perceivable cultural current through the publication of short stories, novels, poetry, visual artwork, and music.[59]

---

59 The first anthology to appear was *Solarpunk: Histórias ecológicas e fantásticas em um mundo sustentável* (2012) in Brazilian Portuguese (the English translation appeared in 2018 under the title *Solarpunk: Ecological and Fantastical Stories in a Sustainable*

Given that the genre is still in its nascency, its stylistic parameters remain under negotiation among artists and critics. What is safe to say is that, as a counterreaction and simultaneously continuation of the cyberpunk tradition, solarpunk is a decidedly technophilic genre. It too is captivated by technoscientific developments and draws on technological metaphors to explore the capacities of human consciousness. But while cyberpunk is simultaneously pervaded by a technophobic ethos in the sense that it tends to take a pessimistic view of the social impact of new technologies, solarpunk has a confident outlook on what humans can achieve through science.[60] In contrast to the quintessential cyberpunk motto "High-Tech. Low Life," solarpunk envisions paths to a brighter future through a combination of high-tech and low-tech modes including gardening, biking, or sailing. Although one will also encounter fantastical high-tech in solarpunk narratives, its primary purpose is no longer to facilitate disembodiment in cyberspace but rather to help humans harmoniously coexist with their environment.

Another aspect setting solarpunk apart from cyberpunk and its other derivatives including biopunk, steampunk, or dieselpunk is that, rather than extrapolating contemporary trends or imagining alternative histories, its main impetus is towards conceptually suspending the separation between the present and the future. By fusing the existing with the imaginary and the innovative with the well-tried, it evokes the palpability of positive change, cultivating a can-do attitude towards the future to harness it for constructive pragmatic action in the here and now. Blending artistic expression with an activist agenda, it embraces "radical forms of organizing that require neither mass revolution nor democratic reform" (Hudson web). At this point, it is worth stressing that this bottom-up

---

World), followed by the English-language collected volumes *Wings of Renewal: A Solarpunk Dragon Anthology* (2017), *Sunvault: Stories of Solarpunk and Eco-Speculation* (2017), *Glass and Gardens: Solarpunk Summers* (2018), *Glass and Gardens: Solarpunk Winters* (2020), *Multispecies Cities: Solarpunk Urban Futures* (2021), and *Saving the Planet: Solarpunk Stories* (2021). Even as solarpunk usually presents itself as a novel genre, its ancestry is often traced to older speculative fiction, particularly the work of Ursula K. Le Guin, Ernest Callenbach, and Kim Stanley Robinson.

60 Many solarpunk works take a strikingly positive view of bioengineering. This does not only represent a sharp departure from cyberpunk with its cautionary tales about the catastrophic fallout from scientific hubris, but also from the spirit of previous environmental movements which have categorically opposed genetic manipulation, for example the modification of crops by now defunct biotechnology cooperation Monsanto.

political ethos is not necessarily based on ideological hostility towards political instruments, but rather on the rationale that the current sociopolitical landscape is inhospitable to the movement's values:

> As a "punk" ideology, solarpunk must be opposed to the political domination of the old. But we may also have to live with it. We just aren't likely to win the numbers game. Solarpunk's strategy should be to create pockets of progress and imagination within a larger political landscape of decay, deadlock and long emergency. (Hudson web)

In opting for hope even as it fully acknowledges the direness of the present situation, solarpunk rejects a cost-benefit-weighing mentality which posits that contributing practical and intellectual energies is only worthwhile if one can expect a return on investment. Instead, it asserts convivial conservation as an ethical axiom independently of quantifiable success, encouraging small transformative acts in the world as we find it.[61]

Solarpunk's Utopian dystopias: Facing the pain of ecological loss

This affirmative stance has led to a misperception of solarpunk as mere escapism or, as the title of one recent article would have it, "A Sunny Therapy Against the Doomsday Mood" (Arte TV web).[62] Such sanguine characterizations fail to do justice to the more somber tones that pervade the genre. In fact, many solarpunk narratives unfold in postapocalyptic settings where survivors have to endure material scarcity, emotional trauma, and hostile environments.

---

61 The incorporation of vegetation into existing architecture is one manifestation of this "practical utopianism" (Owens). It may take the form of communal gardens or large-scale developments such as the Gardens by the Bay in Singapore, the One Central Park in Sydney, or the Bosco Verticale buildings in Milan, completed in 2012, 2013, and 2014, respectively. Although the impact and long-term viability of such projects remains to be seen, such real-life examples point the way to greener urban spaces as envisaged in solarpunk fiction and visual artwork. Among the most circulated graphic representations of solarpunk are the sun-flooded cityscapes by Japanese graphic artist Teikoku Shônen, also known as Imperial Boy. Sometimes categorized as art nouveau punk, his images merge familiar city architecture with lush greenery, wood, and waterways. Shônen's images at times bear a striking resemblance to the work of Austrian visual artist and architect Friedensreich Hundertwasser (1928–2000) who pioneered the integration of irregular organic shapes and greenery into urban architecture.
62 The German-language original is titled "Solarpunk: Die sonnige Therapie gegen die Endzeitstimmung."

Take, for instance, A. C. Wise's short story "A Catalogue of Sunlight at the End of the World." Rendered in the voice of an unnamed elderly botanist, it is a wistful retrospection on his life as his children and grandchildren are about to abandon Earth in the hopes of establishing a new civilization on a distant planet. Addressing his thoughts to his diseased wife, he reminisces how the two of them had dedicated their entire lives to ecological preservation, but to no avail. "We were the last generation who could have turned the tide," he reflects. "By the time [our children] came along, it was too late to undo the mess we'd made. Climate change had already passed the tipping point, and whatever measures we put into place from then on out could only slow things down, not reverse them." (Wise 228) Witnessing the signs of impending cataclysm all around him, he nevertheless opts to remain on the perishing planet to savor the beauty and community that are still left. An affirmation of human resilience, this story is also a confrontation with the prospect of devastating defeat, an elegy to the wonders of an irretrievably lost world.

Another example for solarpunk's oscillation between hopefulness and anguish is Tyler Young's short story "Last Chance," which describes a dystopian scenario in the aftermath of a global ecological catastrophe. To ensure humanity's survival, children are raised underground to protect them from the contamination and blistering temperatures on the surface of the planet. In this claustrophobic subterranean world, non-human life is absent, resources are scarce, the air is stale, and personal privacy non-existent. After graduating school at age sixteen, young people are sent off to work outside under the most severe of conditions in order to sustain the community. The story is narrated from the perspective of teacher Grace, who oversees the children's education. During yet another graduation ceremony, she explains to the parting students that

> Millennia ago, humanity fled the smoking ruin of Earth. [...] Earth's survivors recognized the new planet they found for what it was and named it Salvation. And for a few generations, they treated the planet accordingly. But when the ruin of Earth had faded from living memory [...], the colonists started to make exceptions, to take shortcuts for profit and comfort. And in the span of four hundred years, they destroyed the second Eden. *It can never happen again*, students. There are no more habitable planets within our reach. That's why our forefathers named this planet, your home, Last Chance. (Young 95)

But then Grace goes on to reveal to her audience that all they had been taught about the deadly conditions above had been a lie, and that they actually live on a beautiful, healthy planet. When the students are outraged over having been submitted to a torturous existence without a compelling reason, she insists that "all we've done is justified – it's necessary. You'll never forget what a blessing this world is. You will live modestly, sustainably, and you'll never even consider risking this planet – for anything." (Young 96) The story's message is pessimistic and optimistic at the same time. Pessimistic, because it suggests that humanity has to learn the hard way and that collective wellbeing has to be secured at the price of individual sacrifice. But it is also optimistic in its affirmation that humanity has the capacity to overcome its shortcomings.

By contrast, Julia K. Patt's "Caught Root" asserts that there is yet a future for humanity on planet Earth. Its plot is set in a utopian community called New-Ur, a green haven comprised of "lush gardens and communal spaces, each section self-sustaining and yet part of the greater whole" (Patt 5). As suggested by its name, this society embraces old ways of living as a pathway into the future. Seeking to coexist harmoniously with the natural environment, New-Ur's inhabitants utilize only a minimum of high-tech while still enjoying materially comfortable, spiritually fulfilling, and intellectually productive lives. But even this decidedly hopeful scenario is pervaded by a sense of precarity; far from being the land of plenty, New-Ur's survival hinges on sparing use of resources, disciplined teamwork, and drastically scaled-back expectations regarding the ease and speed of everyday tasks, including travelling. Moreover, we are repeatedly reminded that the community endures only thanks to the ongoing genetic modification of plants which renders them resistant to the altering climatic conditions.

"Caught Root" is above all a bid for the fusion of traditional knowledge with modern science. It is epitomized by the love that develops between the two protagonists, Khadir, the leader of New-Ur, and Ewan, a visiting delegate from another sustainable community called Hillside. In contrast to New-Ur, Hillside relies primarily on new technologies, so that Khadir is skeptical when Ewan suggests a collaboration between both settlements. "Hillside doesn't care about searching our past for solutions," he objects. "It's all new, new, new. The shinier the better." (Patt 5) But his growing feelings for Ewan make him rethink this dismissive stance, and ultimately he agrees to the proposed partnership, clearly persuaded by Ewan's argument that "we might appear to be opposites, but we share the same goals." (5)

As these samples may show, solarpunk fiction comes in varying mixtures of "sunny therapy" with bitter medicine, enticing the reader with the beauty of nature as it confronts them with the ecological losses wrought and experienced by humans. One may object that the genre-typical confidence in the power of science, which at times harkens back uncomfortably to the naïve technological optimism of the 1950s, serves to gloss over rather than to press home the gravity of our ecological predicament. But then again, the same can be said of the numbing pessimism that has become so endemic in speculative fiction. At the very least, solarpunk is exploring new imaginative ground and, maybe most importantly, providing a much-needed venue for the public articulation of environment-related feelings of anguish, sadness, and guilt. Especially given the proliferation of psychological mass phenomena like eco melancholia and meteoranxiety on the one hand and eco-fatigue, ecophobia, and climate change denial on the other, the significance of cultural expressions of grief has increasingly shifted into the public and academic focus as a precondition for meaningful environmental reform. In this context, environmental educator Elizabeth Andre has wondered whether "people won't deal with the climate issue until they have a way to deal with the emotions that come along with it" (Andre 2). Relatedly, in one of the most stirring appeals to break the cycle of complacency, destruction, and dissociation, journalist Jo Confino has argued in a much-noticed article that

> our failure to deal with the collective and individual pain generated as a result of our destructive economic system is blocking us from reaching out for the solutions that can help us to find another direction. ... The question we should all be asking is why aren't we on the floor doubled up in pain at our capacity for industrial scale genocide of the world's species. ... The answer is obvious. We don't need more scientific data or superficial behaviour change initiatives but to engage individuals at a deep emotional, psychological and spiritual level. (Confino, *The Guardian*)

A similar sentiment has been expressed by environmental philosopher Mick Smith who has submitted that a transformation of general environmental attitudes can only occur if we successfully convey "our *feelings* for the natural world in a way that speaks for that world's conservation" (in Barr 202, emphasis original). Solarpunk is particularly suited for tapping into this emotional potential. Unlike cyberpunk, which typically criticizes the exploitation of nature through its very absence – and thus arguably mirrors the psychocultural reflex to repress –, solarpunk tackles this uncomfortable subject head on. Yet its main impulse is to inspire rather than to shock. Addressing itself to a public utterly desensitized by a perpetual onslaught of future horror scenarios, it endeavors to jolt it out of its lethargy by delivering an unexpected dose of positivity.

(More-Than-) Human solidarity

This optimistic approach includes promoting a more favorable view on human nature, which is in stark contrast to the endless parade of marauders, zombies, and tyrannic oppressors populating contemporary speculative fiction.[63] Granted,

---

63 Such fictional narratives have real-life consequences. Especially in the U.S., these

especially in the current political climate of ideological strife and polarization, solarpunk's depictions of communal harmony may seem like quixotic delusions. But even if so, this would hardly count as a disqualifying factor in the context of a literary mode whose very *raison d'être* lies in pushing the boundaries of what is perceived as realistic. Besides, such optimistic visions are actually validated by prominent voices of the scientific community. One of them is historian Rutger Bregman who pushes back against the widespread notion that humans are inherently depraved, selfish, and doomed to failure. While he does not deny that there is much darkness in human history, he points out that the overall trajectory of humanity gives ample cause for optimism. As he writes in *Utopia for Realists* (2017),

> In the past, everything was worse. For roughly 99% of the world's history, 99% of humanity was poor, hungry, dirty, afraid, stupid, sick, and ugly. ... But in the last 200 years, all of that has changed. In just a fraction of the time that our species has clocked on this planet, billions of us are suddenly rich, well nourished, clean, safe, smart, healthy, and occasionally even beautiful (1).[64]

Pointing to this encouraging track record, Bregman goes on to argue that the cultural entrenchment of cynical attitudes towards humanity is based less on empirical evidence than on mental bias towards the negative. In this context, he specifically underlines the suggestive power of fictional narratives and observes that few texts have been more influential in this regard that William Golding's Nobel-Prize winning novel *Lord of the Flies* (1954), which revolves around a group of young boys who get stranded on a lonely island after surviving a plane crash. Out of reach of civilization and adult authority, their life together soon devolves into chaos and deadly violence. Taught to children all over the world, *Lord of the Flies* is often cited as a quintessential parable for the human condition. And yet, Bregman argues, it not only goes against modern scientific insights, but also

---

troubling forecasts have given rise to a large "prepper" scene whose adherents stock survival gear and guns in constant anticipation of imminent deadly violence.

64 It does not escape me that this ameliorative trend has occurred through large-scale industrialization under mostly capitalist modes of production. This, however, does not rescind the fact that the progressing exhaustion of our planet combined with the accumulating effects of economic inequality has rendered the current system ecologically and socially untenable. I thus wish to be clear that I do not deny the advances made in the context of capitalist economies, but rather admit the solarpunk premise that collective progress and "rich, well nourished, clean, safe, smart, healthy, and occasionally even beautiful" people are also conceivable outside the paradigm of endless growth and runaway consumerism.

clashes with practical experience. As counterevidence, he relates an almost identical incident that occurred in 1965: Following a storm, a group of six boys aged thirteen to sixteen were shipwrecked for more than a year on the deserted island of 'Ata, a part of the Tonga archipelago. Bregman makes a point of stressing that it was "[n]ot a tropical paradise with waving palm trees and sandy beaches, but a hulking mass of rock, jutting up more than a thousand feet out of the ocean" (*Humankind* 31). When the boys were eventually discovered in 1966, they

> had set up a small commune with food garden, hollowed-out tree trunks to store rainwater, a gymnasium with curious weights, a badminton court, chicken pens and permanent fire, all from handiwork, an old knife blade and much determination. … The kids agreed to work in teams of two, drawing up a strict roster for garden, kitchen and guard duty. Sometimes they quarreled, but whenever that happened they solved it by imposing a time-out. (*Humankind* 32–33)

The boys stuck together through a series of life-threatening crises. When one of them fell down a cliff and broke his leg, the others "picked their way down after him and then helped him back up to the top. They set his leg using sticks and leaves" (*Humankind* 33). This real-life account of human resilience and cooperation bears a striking resemblance to many of the self-sustaining communities envisioned in solarpunk fiction. The fact the Tongan castaways' story remains an obscure historic episode while *Lord of the Flies* is one of the best-known narratives of our time may serve to underline the claim that we need "alternatives to apocalyptic narratives in which shrinking resources inevitably lead to societal unraveling, mass violence, and the dissolution of trust between peoples" (Botta web). Remarkably, the idea of existence as a constant state of rivalry is also being critically reevaluated in the natural sciences, where "[t]he relatively recent discovery of immense associations of microfungi and flowering plants in symbiotic relationship to each other in ecosystems all over the world has already overturned the dominance of the view of life as solely founded on competitive struggle between species" (Albrecht 305).

Such new perspectives undermine the long-established account of capitalism as a system dictated by nature itself. Instead, they lend credence to solarpunk's recentering on community, cooperation, and symbiosis as an ideological recoil from the capitalist celebration of atomistic individualization, including its solipsistic notions of success. This too is in diametrical opposition to cyberpunk's typical cast of lone-wolf hackers who use their technological expertise to get an advantage over others in a dog-eat-dog world. Solarpunk conversely draws attention to our shared vulnerability and the importance of communal bonds. Crucially, this spirit of solidarity is also extended to the non-human realm, with numerous solarpunk works elaborating on the meaningfulness of relationships with other lifeforms. In fact, an entire anthology titled *Multispecies Cities: Solarpunk Urban Futures* (2021) has been dedicated to the question of how to create well-being not only for humans but also for this planet's other inhabitants. "Plants, animals, fresh air, and clean water are seen only as resources that need to be managed correctly [in contemporary natural science research]," reads the introduction to the volume. "As a result, it is easy to forget the ecological basics: species in the web of life depend on each other in complex ways …. No happiness is to be found on an Earth devoid of more-than-human life." (Rupprecht et al. 3)

This represents a drastic paradigm shift in the way humans perceive of themselves within the biospheric community. Once more, there are some comparisons to be drawn with cyberpunk, and particularly its fascination with the intrusion of high-tech into human bodies and minds. As cyberpunk pioneer Bruce Sterling famously observed in 1986, the digital age had turned technology from something to be merely utilized by humans into something "pervasive, utterly intimate. Not outside us, but next to us. Under our skin; often, inside our minds" (Sterling xiii). In a reversal of this sense of intrusion, solarpunk's impetus goes outwards, calling for a radical expansion of our perimeters of belonging. To repurpose Sterling's words, in solarpunk "nature" is no longer understood as something external to our selfhood, but a pervasive, utterly intimate part of who we are.[65] If, as sociologist John Bellamy Foster has stated, "alienation from the earth is the *sine qua non* of the capitalist system" (Foster 174), this internalization of the non-human could indeed be the decisive step towards a new eco-consciousness.

Conclusion: Capitalist realism versus realist Utopia

We live in a time when the slogan "No Future," adopted half a century ago by the early punk movement, has turned from a sociocritical metaphor to a literal contingency. But solarpunk, the latest incarnation of punk ideology, has undertaken

---

65 This sea change in speculative fiction has parallels in academic literature. Albrecht, for instance, has called for the abolishment of the word "environment" in ecological discourses on the grounds that it reinforces the mental image of human bodies as separate entities in what is in fact an inextricably intertwined symbiotic reality. He has coined the term "symbiocene" to cultivate the mental concept of interrelation rather than separation, thus counteracting the derealization of the non-human as a precondition for its relentless objectification and commodification.

to reclaim the future through radical hopefulness. Granted, this is often a calculated rather than a visceral kind of optimism. As Boffa has remarked rather soberly, the positivity often consists in the mere "fact that these stories envision any future at all." (Boffa web) Broadly speaking, solarpunk does not evoke rosy futures, it does not bank on the aversion of the crisis thanks to sheer luck or a salvific feat of human ingenuity. Instead, it strategically fosters positivity as the mental fuel that drives environmental action in the face of formidable challenges. Where cyberpunk explored the new horizons of virtual worlds, solarpunk retreats from the tropes of virtuality and disembodiment to remind us of our very real physical dependence on our planet. This also includes the rejection of the simulated reality of capitalism which is prone to sacrificing the quality of immediately experienced life for abstract monetary value, being as it is unable to "take into account values like the quality of human interaction, culture, or the desire for a healthy environment" (Jaffee 19). Solarpunk challenges fundamental capitalist tenets about the nature of success and progress by replacing the imperative of perpetual expansion with restraint, competition with cooperation, and anthropocentric with ecocentric paradigms.

At first glance, solarpunk's optimistic stance may seem like the most improper of responses in a time when humanity is facing "death on the largest scale imaginable – that of life on earth itself" (Orr 181). But then again, it may be overdue to question why our cultural discourses are so predisposed to conflating optimism with naivety, and cynicism with critical thinking. For as real-life experience has shown, neither pessimistic lethargy nor ironic detachment have served us well in devising alternatives to an economic system based on the rapacious exploitation of our planet.

In conclusion, I would like to give the last word to the late Mark Fisher. When over a decade ago he looked at the state of speculative fiction, he somewhat wistfully observed that "[o]nce, dystopian film and novels were exercises in … acts of imagination – the disasters they depicted acting as narrative pretext for the emergence of different ways of living" (Fisher 2). If he thought that no longer the case, he attributed it to "the fact that capitalism has colonized the dreaming life of the population" (Fisher 8). In daring to radically reimagine human civilization, solarpunk provides new topographies for future aspirations – and thus takes a momentous step towards decolonizing the collective mind.

References

Albrecht, Glenn. "Solastalgia and the New Mourning." In *Mourning Nature: Hope at the Heart of Ecological Loss and Grief*. Eds. Ashlee Cunsolo and Karen Landman, Montreal: McGill-Queen's University Press, 2017. 291–315. Print.

Andre, Elizabeth Kathryn. "Journeying Through Despair, Battling for Hope: The Experience of One Environmental Educator." Doctoral Dissertation at the Faculty of the Graduate School of the University of Minnesota, 2011. Web.

Arte TV. "Solarpunk: Die sonnige Therapie gegen Endzeitstimmung." Web.

Barr, Jessica Marion. "Auguries of Elegy: The Art and Ethics of Ecological Grieving." In *Mourning Nature: Hope at the Heart of Ecological Loss and Grief*. Eds. Ashlee Cunsolo and Karen Landman, Montreal: McGill-Queen's University Press, 2017. 190–226. Print.

Boffa, Adam. "At the Very Least We Know the End of the World Will Have a Bright Side." *Longreads*. Web.

Bregman, Rutger. *Humankind: A Hopeful History*. London: Bloomsbury, 2021. Print.

Bregman, Rutger. *Utopia for Realists: And How We Can Get There*. London: Bloomsbury, 2017. Print.

Confino, Jo. "Grieving Could Offer a Pathway Out of a Destructive Economic System." *The Guardian*, 2 October 2014. Web.

Fisher, Mark. *Capitalist Realism: Is There No Alternative?* Portland: Zero Books, 2009. Print.

Flisfeder, Matthew. "Blade Runner 2049 (Case Study)." In *The Routledge Companion to Cyberpunk Culture*. New York: Routledge, 2020. 144–150. Print.

Foster, John Bellamy. *Marx's Ecology: Materialism and Nature*. New York: Monthly Review Press, 2000. Print.

Hickman, Caroline et al. "Climate Anxiety in Children and Young People and Their Beliefs About Government Responses to Climate Change: A Global Survey." *The Lancet*, December 2021. Web.

Hudson, Andrew Dana. "On the Political Dimensions of Solarpunk." *Medium*. Web.

Jaffee, Daniel. *Brewing Justice: Fair Trade Coffee, Sustainability, and Survival*. Berkeley: University of California Press, 2007. Print.

Kretz, Lisa. "Emotional Solidarity: Ecological Emotional Outlaws Mourning Environmental Loss and Empowering Positive Change." In *Mourning Nature: Hope at the Heart of Ecological Loss and Grief*. Eds. Ashlee Cunsolo and Karen Landman, Montreal: McGill-Queen's University Press, 2017. 258–291. Print.

Kretz, Lisa. "Hope in Environmental Philosophy." *Journal of Agricultural and Environmental Ethics* 26, 925–944 (2013). Web.

Mark, Andrew and Amanda Di Battista. "Making Loss the Centre: Podcasting Our Environmental Grief." In *Mourning Nature: Hope at the Heart of Ecological Loss and Grief*. Eds. Ashlee Cunsolo and Karen Landman, Montreal: McGill-Queen's University Press, 2017. 227–257. Print.

Orr, David W. "The Uses of Prophecy." In *The Essential Agrarian Reader: The Future of Culture, Community, and the Land*. Ed. Norman Wirzba, Washington: Shoemaker and Hoard, 2003. 171–187. Print.

Owens, Connor. "What is Solarpunk?" *Solarpunk Anarchist*. Web.

Patt, Julia K. "Caught Root." In *Glass and Gardens: Solarpunk Summers*. Ed. Sarena Ulibarri, Albuquerque: World Weaver Press, 2018. 3–9.

Republic of Bees. "From Steampunk to Solarpunk." Web.

Rupprecht, Christoph et al. "Introduction." In *Multispecies Cities: Solarpunk Urban Futures*. Albuquerque: World Weaver Press, 2021. 1–10. Print.

Sargent, Lyman Tower. *Utopianism: A Very Short Introduction*. Oxford: Oxford University Press, 2010. Print.

Springett, Jay. "Solarpunk: A Refernce Guide." *Medium*. Web.

Sterling, Bruce, "Preface." In *Mirrorshades: The Cyberpunk Anthology*. Ed. Bruce Sterling, New York: Arbor House, 1986. ix–xvi. Print.

Wise, A. C. "A Catalogue of Sunlight at the End of the World." In *Sunvault: Stories of Solarpunk and Eco-Speculation*. Eds. Phoebe Wagner and Brontë Christopher Wieland, Upper Rubber Boot Books, 2017. 221–236. Print.

Florian Andrei Vlad

Rednecks Gone Wild: Ecodefense and Edward Abbey's Monkeywrenching

Abstract: An examination of Edward Abbey's work invites one to see the author exploring the "eco imperatives," gradually moving away from anthropocentrist to biocentrist positions that have gained prominence in ecocritical studies over the last few decades, with touches of anarchism announcing his monkey wrenching approach in the novel that this essay focuses upon. The militant dimension evoked by the monkey wrench is to be linked to the almost equally "luddite" term of "ecodefense."

Introduction

One can approach Edward Abbey's *The Monkey Wrench Gang* (1975) from a variety of perspectives, more or less militant or radical, anarchist or otherwise, especially if one starts from the name of the impromptu gang that causes trouble to technology, consumerism, capitalism and other *-isms* that are seen to threaten the environment, notably the wilderness of the great outdoors of the American Southwest. The name of the monkey wrench gang, immediately linked to the figure in whose memory Abbey wrote his best-known novel, Ned Ludd, the first frame-breaker of the British Industrial Revolution, conjures up the meaning of the idiom "to put a monkey wrench into the works," synonymous with sabotage. Nevertheless, before considering the "Luddite" dimension of Abbey's work, a more "peaceful" generic label comes to mind.

A twenty-year-old "coming from an Appalachian hillbilly background and with a poor choice of parents," as he styles himself (Abbey 1988: 9), Abbey reached the desert Southwest and fell in love with the rugged beauty of the wilderness there, a beauty that he would poetically deal with in his *Desert Solitaire: A Season in the Wilderness* almost two decades later. In that book, he defined his new privileged relationship, which lasted much more than a season. Aware that each person carries in their heart an idealized place as home of choice rather than of origin, Abbey makes his own statement to that effect: "For myself, I'll take Moab, Utah. I don't mean the town itself, of course, but the country which surrounds it – the canyonlands. The slickrock desert. The red dust and the burnt cliffs and the lonely sky – all that which lies beyond the end of the roads" (Abbey 1971: 1). What would later be seen as the "monkey-wrenching author" was, first and foremost, a nature writer, in the opinion of a number of notable critics.

Ecocriticism, as surveyed by Greg Garrard (2004), deals with a number of key issues and tropes, among which pollution (1–15), pastoral (33–58), wilderness (59–84) and apocalypse (85–107) that feature prominently in any approach to Edward Abbey's both non-fictional and fictional texts. The ecocritical approach accommodating ecological and environmental concerns had been derived directly from the book proper, although clearly articulated almost two decades later (see Dana Phillips 579). Apocalypticism as an ecocritical trope both responds to, and produces, crises in the opinion of Garrard (2004: 86) and what the monkey wrench gang members do is a good illustration of both these tendencies, as it will be seen, in addition to the other defining characteristics of writing affected and affecting the environmental turn. They both respond to crises and cause ecodefense trouble, from milder to more serious acts of eco-terrorism.

Cherryl Glotfelty, the well-established ecocritic, dealing with environmental literature, notes the prominence of such dimensions as writing concerned with natural history, idiosyncratic responses to nature, philosophical assessments and interpretations of nature. She examines such critics of nature writings as Don Scheese, placing Edward Abbey in a tradition going back to Thoreau meant to "instill a land ethic" in the audience:

> Don Scheese's "Desert Solitaire: Counter Friction to the Machine in the Garden" considers one of Thoreau's most colorful followers, Edward Abbey. Scheese insists that although Abbey resisted the label "nature writer," he nevertheless falls squarely in the tradition of nature writing established by Thoreau and carried on by John Muir and Aldo Leopold, all of whom sought to instill a land ethic in the American public. (Glotfelty xxxi)

Don Scheese's title above obviously evokes the name of Leo Marx, more specifically his *The Machine in the Garden: Technology and the Pastoral Ideal in America*. The by now classic 1964 volume attempting to capture the spirit of America ever since its "idyllic" beginnings, like any other important book, elicits both approval and criticism, and ecocriticism is obviously one such response. In his book, Marx (the peaceful, "pastoral one," not the instigator of Communist revolutions) makes a claim that can be taken as a starting point in any debate of the complex, problematic relationship between the apparently different realms of Nature and Science and Technology: "The pastoral ideal has been used to define the meaning of America ever since the age of discovery, and it has not yet lost its hold upon the native imagination" (Marx 3).

Coming back to the same Don Scheese, throughout the eight chapters of his *Nature Writing: The Pastoral Impulse in America*, the author gives an outline and defines the key features of a nature writing genre touched by what he considers the pastoral impulse. Among the most representative authors of this genre

Abbey appears, once again in the wake of Thoreau's work, in the company of such authors as John Muir, Mary Austin, Aldo Leopold and Annie Dillard. In so doing, Scheese examines the competing preoccupations which have featured prominently in the pastoral genre for millennia, every since Theocritus: wilderness vs civilization, anti-modernism vs. progress, biocentrism vs anthropocentrism. Among these isms, Bret Olsen also challenges one to consider the relation between myth and environmentalism, with the latter as a rejection of the myth of the West, a peculiar stance that many nature writers, Abbey included, are to cope with:

> Myth's role in shaping western reality presents a visible bevy of seemingly comical, yet potentially disastrous ironies. Witness the ultimately self-destructive practice of land developers plowing up miles of desert to plant lawns and golf courses. Certainly profit margins provide the ultimate pull, but the seemingly vast abundance of their monumental natural surroundings often obscures the reality of a finite water supply. (Olsen 124)

The apparently serene dimension of the pastoral impulse that Edward Abbey shares with a long tradition of nature writers acquires a distinctly rebellious, anarchist quality, especially in association with his illustrious American predecessor, Thoreau. Abbey will emulate him in real life with his civil disobedience tendencies, thus gaining a prominent position in the FBI files on dangerous individuals. In a conversation recorded by good friend and fellow nature writer Jack Loeffler, included in "Edward Abbey, Anarchism and the Environment," the author of *The Monkey Wrench Gang* defines himself as a "registered anarchist," as a "barefoot wilderness eco-freak anarchist" in a tradition bringing together both Thoreau and the Russian Peter Kropotkin, the advocate of anarcho-communism: "In the realm of ideal politics, I'm some sort of an agrarian, barefoot wilderness eco-freak anarchist. One of my favorite thinkers is Prince Kropotkin. Another is Henry Thoreau" (Loeffler 43).

However, if one examines the competing impulses in Abbey's work as a whole, *Desert Solitaire: A Season in the Wilderness* (1968), an autobiographical text preceding the registered anarchism of *The Monkey Wrench Gang*, he is closer to William Wordsworth in a more solitary and contemplative mode in the same tradition of nature writing. The book recounts, in a non-fiction mode that is suffused with undeniable poetic power, Abbey's beginning of a lifelong relationship with the monumental natural architecture of the Southwest when he began his initially contemplative "environmentalism" as a park ranger in what is now the Arches National Park. Significantly, that came at a time when American authorities began to pay particular attention to the wilderness.[66] While reviewing the

---

66 See The Wilderness Act of 1964 (justice.gov) for more information on this issue.

beginnings of the deep ecology movement of the 1960s, of which Abbey will be one of its prominent inspirational figures, George Sessions acknowledges and supports a renewed interest of the authorities at that time, while also mentioning previous initiatives:

> Given the state of environmental deterioration by the early 1960s, the administration of John F. Kennedy was about to launch the third major conservation effort of the century (the first two occurred during the administrations of the two Roosevelts). Secretary of the Interior Stewart Udall signaled that effort with the publication of The Quiet Crisis. (Sessions 105)

Unfortunately, many environmentalists then began to realize the emergence of an opposite, anti-environmentalist trend, supporting the big business interests intent on intruding into the unspoilt beauty of some of the scenic sites of the American sublime, so to speak. That largely explains the backlash that the most extreme forms of ecodefense will assume in those years, and again Abbey's stance as a registered "eco-freak anarchist" appears to justify itself.

What influence the daffodils of the tame, picturesque Lake District had on the British Romantic poet was replaced, as far as Abbey is concerned, by the influence of the vast expanses of desert and the sublime abysses of the canyons of the American Southwest. In what follows, it is the eco-freak anarchist that will be the main focus, while the whole of Abbey's complex vision to the environment, as well as the various trends and tradition in relations to which his work gains particular significance, as well as the poetry of the arid lands in Desert Solitaire, will complete the comprehensive landscape of one author and the multiplicity of the environmental discourses more or less contemplative, more or less "anarcho-terrorist."

An examination of Edward Abbey's work will invite one to see him exploring the "eco imperatives," gradually moving away from anthropocentrism to biocentrism positions that have gained prominence in ecocritical studies over the last few decades, with touches of anarchism announcing his monkey wrenching approach in the novel this article focuses on. The militant dimension evoked by the monkey wrench is to be linked to the almost equally "luddite" term of "ecodefense."

Ecodefense is a term first coined by the (in)famous *Ecodefense: A Field Guide to Monkeywrenching* (1985), edited by Dave Foreman, with a foreword by Edward Abbey, subsequently re-published several times in both physical and virtual *samizdat*. Jeremy Lloyd, in his interview with Foreman, the co-founder of *Earth First!*, uses the phrase "rednecks for wilderness" in a December 2005 article in the *The Sun Magazine* (as distinct from The Sun) to refer to both Foreman and to other *champions of conservation of the 20th century with whom he also identifies (Lloyd web). They are more than conservationists, they are eco-defenders.*

Drawing inspiration from a radical green literary tradition dominated by Abbey's *The Monkey Wrench Gang*, ecodefense is a framework of environmental thought and action that is biocentric (thus eschewing the anthropocentric utilitarianism of mainstream environmentalism), militant (favoring direct action over traditional politics), populist (eschewing cosmopolitan liberal pieties) and neo-Luddite (opposed to the excesses of technology and industrialization). Its methods have been used by leaderless activist groups such as Earth First!, Earth Liberation Front, Animal Liberation Front, and the Sea Shepherd Conservation Society. While there is no consensus on using a unitary term to encompass all environmentalist activism influenced by Abbey, Foreman, or Earth First!, the term ecodefense is used here to refer to a particular cluster of militant, anti-technocratic intellectual and activist traditions which will advocate direct action and civil disobedience to oppose the destruction of natural habitats.

Ecodefense, ecodefender, as well as *eco-anarchism*, have become established in contemporary culture, and John Algeo and Adele Algeo give the words attention as early as 1988. They do not fail to trace ecodefense and ecodefender to the above-mentioned *Ecodefense: A Field Guide to Monkeywrenching* (Algeo & Algeo 347). While ecodefense might not be a specific political philosophy, the term itself highlights the importance of action ("defense") which distinguishes it from the more generic terms "environmentalism" and "conservation." All these trends and traditions deserve more detailed consideration before dealing with Edward Abbey's achievement, thus placing the "monkeywrenching masterpiece" in a proper, comprehensive framework. This move apparently goes counter to Ned Ludd's frame breaking revolt, obviously meant to supply frames in relation to which the author of *The Monkey Wrench Gang* is to be understood and assessed, "framed," so to speak, if one gives the word a positive spin.

Broadening the conceptual and theoretical frames of Edward Abbey's environmentalist monkey-wrenching

The illustration on the dust jacket of the tenth anniversary edition (1985) of Edward Abbey's volume makes the uninformed reader think that this is an adventure book for children only, probably a graphic book. It turns out that it is not a graphic book, also enriched by a number of additional illustrations in its body. Good books have many uses, addressing a variety of audiences. The introductory lines above have already defined specific areas in which the text engages in acts of environmental civil disobedience, but the book's revolutionary impact should nevertheless be seen in the long (largely pastoral, as already mentioned) tradition of nature writing, ranging from conservationism to environmentalism, but going all the way to eco-terrorism, thus largely defining the persona non grata status that the secret services and honorable, law-abiding citizens of the 1960s and 1970s associated with the author as troublemaker. It is worth noting that today, away from the more exuberant, countercultural ethos of the baby-boomers of the sixties and seventies, eco-*terrorism* evokes the T word which has acquired the sinister post-9/11 connotations that the French Revolution and its Reign of Terror had made acceptable for a long tradition of militants worldwide. It is also worth noting that, post-9/11, in the wake of both literary "monkeywrenching" texts as Abbey's and of extremist, militant, "extra-literary" approaches to the sensitive issue of the endangered environment calling for help, the more anthropocentric, more specific threat to human lives in society has led to the conflation of more general environmental concerns with closer-to-home-sweet-home concerns for individual and group security. This has been reflected in forms of scholarship bringing together terrorism, security and environmentalism, and undergoing a dramatic boom that is explained by a renewed interest in their combined problematic, as Loadenthal notes:

> In the wake of the 9/11 attacks, the often-linked fields of Terrorism Studies and Security Studies have witnessed a boom, accompanied by the more general rise in university studies directed at Islam, political Islam, terrorism and Middle Eastern politics[1]. Subsequently, new approaches have been developed within a host of "critical" fields, including Critical Terrorism Studies [2] and Critical Security Studies. (Loadenthal 16)

Angela Mertig, in her informative "From Conservationists to Environmentalists: The American Environmental Movement" (2015), notes how the American environmental movement has gone through significant developments since the conservation movement at the end of the 19th century. She is aware that it has shown an evolution from the limited preoccupation with the preservation of regional resources and the conservation of picturesque sites, forests and parks toward the wider global preoccupations with pollution and the prevention of climate change.

The first militants and their organizations defined their activity as conservation and preservation. The concern with and about conservation and preservation are still there all right, but the ideology has broadened to include a framework which has been better defined as ecologist or environmentalist (Mertig 55). The author concludes her survey by contrasting the motto of the Sierra Club in 1892 and in 2014. If the conservationist users of the initial motto contented themselves "to explore, enjoy, and render accessible the mountain regions of the Pacific Coast," the contemporary environmentalists of the same club aim to "Explore, Enjoy and Protect the Planet" (Mertig 75). The well-documented and informative essay thus provides a "peaceful," general context in which a variety of environmental movements can be integrated, brandishing or not such a peculiar weapon as a monkey wrench, of all things, in their defense of the natural environment.

The militant, "luddite" dimension of ecodefense can more specifically be linked to one particular thread of the general movement from conservationism to radical environmentalism and more or less "monkeywrenching" eco-terrorism, and Horacio R. Trujillo's "The Radical Environmentalist Movement" undertakes to provide the general frame in which such positions as Edward Abbey's can better be understood and interpreted. He claims in his text that the FBI considers The Earth Liberation Front (ELF), the Animal Liberation Front (ALF), and some other divisions of the radical environmentalist movement to represent America's top terrorism threat (Trujillo 141). Nicole A. Tishler's 2018 article, "Fake Terrorism: Examining Terrorist Groups' Resort to Hoaxing as a Mode of Attack," completes the general picture as sketched by Trujillo on the authority of the FBI. In relation to fictional accounts of ecodefense and eco-terrorism as those imagined in Abbey's novel, but in close connection with goals and attitudes of real militants in real places, fake terrorism and its hoaxes are also relevant to *The Monkey Wrench Gang,* in keeping with the less violent ideologies of both mainstream environmentalism and animal rights movements:

> Animal rights and environmentalist ideologies typically involve strong norms of non-violence, manifested in groups' explicit calls to protect human life. This orientation would logically extend to anti-abortionist groups. For groups espousing these motivations, hoaxes may serve as a morally-consistent means of publicizing their cause, spreading fear, and draining responder resource. (Tishler 5)

Abbey's fictional account of his eco-terrorist gang does not appear to spread fear, the "revolutionary" behavior of what may appear to be a group of rednecks having fun in the great outdoors of the American Southwest being a far cry from the terrible deeds of the Glanton gang in the same area one century before in Cormac McCarthy's *Blood Meridian*. Abbey's characters roam the environment of the desert, moving in the ideological realms between fake terrorism and eco-terrorism, never settling down to one particular position, between the rough and tough anarchist eco-warriors and the reflective, pensive romantic rednecks in love with the great outdoors of the sun-drenched stony Southwest.

The radical pastoral and the machine in the American Southwest

In his book, *The Rights of Nature: A History of Environmental Ethics*, Roderick Frazier Nash recalls Dave Foreman in 1981 instigating the passage from the Thoreau-type of non-violent forms of civic disobedience to violent forms of eco-defense. Foreman "urged his colleagues to 'free shackled rivers' and remarked that 'the finest fantasy of eco-warriors in the West is the destruction of [Glen Canyon] Dam and the liberation of the Colorado [River.]'" (Nash 167). This eco-terrorist invitation in 1981 is reminiscent both of heated debates about that dam and about the beginning of Abbey's novel (Prologue: The Aftermath), in which is imagined the more-than-monkeywrenching liberation of the "tame and domesticated" Colorado from the steel and concrete clutches of the Glen Canyon Dam on the occasion of the inauguration of the bridge close to it. The bridge is imagined as blowing up, while a terrorized governor, a highway commissioner and two high-ranking officers of the Department of Public Safety know what next lies in store: the eventual destruction of the Glen Canyon Dam itself.

It is the first fantasy of eco-terrorism in the novel reminiscent of what John R. Knott calls by the relatively nice and peaceful phrase of "the romance of wilderness" to refer to Abbey's radical environmental attitudes:

> Anyone who knows Abbey's writing will recognize one kind of extremism in his assaults on proprieties of any kind he can imagine and his fantasies of subverting the ranchers, developers, dam builders and other enemies of wilderness whose ways he never tired of ridiculing. (Knott 331)

The four ecodefense musketeers, in this sense, are also romantics of the wilderness. These eco-terrorist characters appear to illustrate the problematics of Edward Abbey's MA thesis at the University of New Mexico, "Anarchism and the Morality of Violence." The ethics contemplated by the radical environmental ideology that George W. Hayduke and his accomplices espouse move away from an anthropocentric one toward ecocentrism. It is not only promoting animal liberation, but giving equal attention and respect to the inorganic, to rivers, oceans, mountains threatened by contemporary environment exploitation by big business interests. Abbey is quoted as saying that environmentalism promoted "a recognition... of the right of nonliving things – boulders, for example or an entire mountain – to be left in peace" (Nash 168). If the magnificent four eco-warriors will sometimes hesitate about blowing up dams and liberating tamed rivers from them, they will content themselves with more modest targets.

Dealing with the first type of these modest targets is illustrated by the first of the four characters. This is A.K. Sarvis, M.D., in the company of his female assistant, Ms. Bonnie Abzug. Does the name and the company foreshadow some sort of Bonnie and Clyde criminal relationship? The fact is that eco-terrorism as ecodefense in the novel begins with a far from grand gesture from the new Bonnie and Dr. Sarvis duo. The more or less respectable Dr. Sarvis, Albuquerque physician by day, has one particular eco-defensive hobby while medically off-duty. He occasionally sets fire to billboards along the highway – U.S. 66 which is to be swallowed by the larger interstate freeway. With a big can of gasoline he generously sprinkles the legs and support members of the billboard, then sets it ablaze. The reason he does this is most likely due to the association of billboards in particular, advertising in general, with the promotion of consumerism, one of the mortal sins of affluent capitalism. Political ads are also among the possible targets, such as WHAT'S WRONG WITH BEING RIGHT? JOIN THE JOHN BIRCH SOCIETY! The bumper stickers on A.K. Sarvis's car proclaim, I AM PROUD TO BE AN ARMENIAN!, which might sound flippant, and a more earnest, conservationist, environmentalist GOD BLESS AMERICA. LET'S SAVE SOME OF IT (Abbey1985: 21–22).

The next of the group to be shown will benefit from a less farcical introduction, and from a very impressive name, devoid of the criminal connotations of that of Ms. Bonnie Abzug: George Washington Hayduke. He had served in Vietnam in the Special Forces. After fighting for two years in the jungle and spending one year as a POW of the Vietcong, he had got back to his beloved American Southwest to discover that it had changed beyond recognition. Back from Vietnam, the character looks at the city of Tucson, which is now surrounded by a series of Titan Intercontinental Ballistic Missile bases, while the desert is scraped bare of its vegetation, of its wilderness life, by huge bulldozers that remind him, a Vietnam veteran, of the American plows leveling Vietnam. The wilderness is being replaced by machine-like wastes with big companies having the lion's share in the massive apocalyptic destruction:

> These machine-made wastes grew up in tumbleweed and real-estate development, a squalid plague of future slums constructed of green two-by-fours, dry-wall fiberboard and prefab roofs that blew off in the first good wind. This in the home of the free creatures: horned toads, desert rats, Gila monsters and coyotes. Even the sky, that dome of delirious blue which he had once thought was out of reach, was becoming a dump for the gaseous garbage of the copper smelters, the filth that Kennecott, Anaconda, Phelps-Dodge and American Smelting & Refining Co. were pumping through stacks into the public sky. A smudge of poisoned air overhung his homeland. (Abbey 1985: 23)

Hayduke's apocalyptic realization of what is going on in Tucson and around it in the once unspoiled desert magnificence is a good example of what Garrard had defined as one of the prominent eco tropes. This explains the character's anger and it provides the motivation for Hayduke and the whole gang to behave the way they do in their subsequent eco-war against the corporate enemies of the environment.

His anger is somewhat eased by the pastoral impulse, by the contemplation of what has not been ruined by the encroachment on the wilderness perpetrated by American culture and civilization of one particular sort. Hayduke opens himself through his pores and nerve ends to the stillness of his beloved Arizona desert, remembering what for others might be inessential or even ugly: "the scrubby little trees: the mesquite [...], the green-barked paloverde with its leafless stems (the chlorophyll is in the bark), the subtle smoke tree floating like a mirage down in the sandy wash (Abbey 1985: 24).

Like Abbey, Hayduke is both a nature lover (in the pastoral tradition) and an anarchist who resents authority, especially when authority abuses its powers. This adds a picaresque touch to the character's overall apocalyptic vision of destruction and counter-destruction. Before any monkeywrenching eco-terrorist plans in a group which is about to take shape before long, Hayduke has to settle his account with a policeman who had once wronged him. Now that he is back from Vietnam, he has to find him and to punish him in keeping with his wrongdoing, with readers most likely expecting some sort of Rambo feat. Hayduke will modestly content himself with stealing the policeman's car and leaving it on the rail tracks in front of an advancing locomotive, not before helping himself to a few souvenirs: a shotgun, a riot helmet and a six-battened flashlight. Apparently, he has to live up to his name, accommodating both George Washington and Hayduke.

Hayduke appears to display a peculiar mixture of identities, a romantic redneck of a troublemaker with an anarchist, eco-terrorist streak, with the contribution of the name of the first American president to be figured out by the reader, if possible. Solitary in the great outdoors, happy as a pig in shit, as the sympathetic narrator sees him, the character first takes a leek, then kneels, takes a pinch of red sand and eats it ("Rich in iron. Good for the gizzard"). The nature lover then stands up facing the river and the soaring cliffs, the glorious sun going down, in an unwitting imitation of young William Wordsworth in conversation with the grandeur of the Alps, everything enriched with a notable dose of American Southwest wildlife:

> He spreads his legs solidly on the rock and lifts his arms wide to the sky, palms up. A great and solemn joy flows through bone, blood, nerve and tissue, through every cell of his body. He raises his head, takes a deep breath – The heron in the canyon, a bighorn ram on the cliff above, one lean coyote on the rim across the river hear the sound of a howl, the song of a wolf, rise in the twilight stillness and spread through the emptiness of the desert evening. One long and prolonged, deep and dangerous, wild archaic howl, rising and rising and rising on the quiet air. (Abbey 1985: 34)

The "wild archaic howl" at the end of the quote above makes one think of one particular construction of tough masculinity in a book from Abbey's generation, Robert Bly. In *Iron John: A Book About Men*, Bly advocates a return to what one might call "innocent, primeval male wilderness," unspoiled by civilization. In his book, Bly laments contemporary American men:

> The warriors inside American men have become weak in recent years, and their weakness contributes to a lack of boundaries, a condition which earlier in this book we spoke of as naivete. A grown man six feet tall will allow another person to cross his boundaries, enter his psychic house, verbally abuse him, carry away his treasures. (Bly 146)

The inner warriors of the monkey wrench gang, both male and female, including Hayduke, will not allow the "enemy" to carry away their treasures. Their treasures, as it turns out, consist of the unspoilt nature of the canyons and of the rugged desert beauty of the American Southwest.

The next to be introduced of the four eco-terrorists, all of them still unaware of their preordained monkey wrench gang membership, is Joseph Fielding Smith. Again, the name carries weight, albeit ironically. Joseph Fielding Smith Sr. and Joseph Fielding Smith Jr. were the sixth and tenth presidents of the Mormon Church. Abbey's character by that name was born into the Church of Jesus Christ of Latter-Day Saints (the Mormon Church). He is said to be on lifetime sabbatical from his religion. What he preserves from the doctrine of his Mormon forebears is a sort of plural marriage that he consistently practices. He has occasional wives from Cedar City to Bountiful to Green River, all wives and places in Utah, the state founded by the Mormons. Joseph Fielding Smith's brief temporary stay in each of the above-mentioned Utah cities with the respective wives explains his acquired *nom de guerre*, so to speak. His Utah wives have renamed him Seldom Seen Smith, under which name he is to be acknowledged by all his acquaintances, including the other characters and the readers themselves.

Seldom Seen Smith has the most specialized job in the canyonlands of the American Southwest of those who will soon make the monkey wrench gang. He is the wilderness guide and the river runner that organizes float trips down the Colorado River from Lee's Ferry to the Grand Canyon. Nevertheless, like

Hayduke and the other monkey wrench characters, he hates the damage that has been done to the environment, and the Glen Canyon Dam is one of these monsters. He calls "the blue death" the large water reservoir walled in by the dam, while remembering the mysterious beauty of the desert rock architecture now hidden by the "dead water" of the new lake. He remembers a "golden river flowing to the sea," as well as many canyons with such names as Hidden Passage, Salvation, Last Chance, Forbidden, Twilight, as well as great amphitheaters called Music Temple and Cathedral in the Desert. They are "slowly disappearing under layers of descending silt" (Abbey 1985: 36).

The Glen Canyon Dam is described by the same Seldom Seen Smith in less poetical, very factual terms, a juggernaut that defies the desert's environment beauty with its colossal amount of cement, a Moloch that had swallowed human lives and a lot of taxpayers' money: "Vital statistics: 792,000 tons of concrete aggregate; cost $750 million and the lives of sixteen (16) workmen. Four years in the making, prime contractor Morrison-Knudsen, Inc., sponsored by U.S. Bureau of Reclamation, courtesy U.S. taxpayers" (Abbey 1985: 34).

There is one thing that Smith claims that he has not tried yet to fight such monsters, reminiscent of his very religious name: praying for a precision earthquake right at the dam, thus adding, in typical Edward Abbey fashion, a little bit of clowning to the seriousness of his radical, but more down to earth environmentalist creed. His prayer to God is long, worth quoting in full but impossible to do so in a short text like this. Smith reminds the Almighty of what the place looked like before "the bastards from Washington" came and ruined it all. He then, in a powerful stream of poetic language reminiscent of Abbey's *Desert Solitaire*, evokes the beauty of the river, of the mountains and of the wild life around the canyons with minute descriptions of specific places and moments of the day and of the year. Short of enough space for the whole prayer, the ending will suffice here, evoking the wry clownish tone of the Mormon on a lifetime sabbatical mentioned above: "Listen, are you listenin' to me? There's somethin' you can do for me, God. How about a little old *pre*-cision-type earthquake right under this dam? Okay? Any time. Right now for instance would suit me fine." Seeing no immediate response from the Almighty, Smith thus ends his invocation: "Okay, God, I see you don't want to do it just now. Well, all right, suit yourself, you're the boss, but we ain't got a hell of a lot of time. Make it pretty soon, goddammit. A-men" (Abbey 1985: 38).

Passages in which Romantic descriptions of what is seen as left of the beauty of American wilderness (remember the text on the bumper sticker, GOD BLESS AMERICA. LET'S SAVE SOME OF IT) will alternate with humorous, irreverent, sometimes unpolitically correct now stretches. The circle of eco-anarchists will soon close with Smith the river runner and guide meeting Dr. Sarvis and Ms. Bonnie Abzug. The latter appears to be a runaway from New York's Bronx canyons, very much like Edward Abbey himself, a native from Pennsylvania coming to his chosen homeland in the Southwest. The former is described as having cherished his billboard-burning ecodefensive hobby for some time now, to such an extent that, like Edward Abbey himself through his eco-anarchist writing, he has already made a big environmental splash. An environmentalist virus appears to have contaminated the public space. Various anonymous letters appear addressed to officials, claiming responsibility for the billboard-burning acts of vandalism, to reach the media that will further publicize, positively or negatively, the deeds and a dramatic name for such radical behavior:

> The newspaper stories mentioned "organized bands of environmental activists," a phrase soon shortened to the much handier and more dramatic "eco-raiders." The county attorneys warned that the perpetrators of these illegal acts, when caught, would be prosecuted to the fullest extent of the law. Nasty letters, pro and con, appeared in the Letters-to-the-Editor columns. (Abbey 1985: 48)

Dr. Sarvis, however, is aware that this billboard-burning business is kid stuff for an eco-raider and that he was meant for more serious things. He knows the cartography of the area, which will turn into Edward Abbey's literary cartography as a whole in the novel, a mapping that the novel shares with that of Desert Solitaire, among other texts. Unlike Faulkner's Yoknapatawpha, Abbey faithfully records real places in this geography of the Four Corners, the intersection of Utah, Colorado, New Mexico, Arizona where the picaresque narrative that follows will unfold. Again, it is the faithful description of the desertscape that a former park ranger knows by heart, from Flagstaff and Tucson in the south to Mollie's Nipple and to Grand Canyon, Lee's Ferry, west to Cedar City, center to Valley of the Gods and Mexican Hat, names of hills and valleys and canyons, as well as more or less godforsaken cities and ghost towns. Seldom Seen Smith is also to be seen as an authority on this space, while Bonnie, coming from the Bronx and Hayduke from Vietnam, are quick to learn the feel of the desert land. They feel the threat and they share the anger. They see big business being responsible for the massive depredation of the wilderness: strip mines, oil companies, logging companies, coal-burning power plants, power lines and railroads are seen as conspiring to deface the most spectacular area of the American Southwest.

What can the four characters do if God does not appear to answer Seldom Seen Smith's peculiar prayer? Largely following the mock-heroic and picaresque tradition, which one can safely say that was announced from the very beginning – after all, Seldom Seen Smith's real name, Joseph Fielding, is also reminiscent of Henry *Fielding* and of his hero, *Joseph* Andrews – the characters engage in acts of eco-terrorism, wrecking bulldozers, heavy tractors and other environmental-unfriendly machines that the previously-mentioned "enemies" are using to deface the area shared by the fictional cartography of the book and the real geography of the place.

Conclusion

In addition to comic delineation of characters, sometimes verging on the grotesque, and the nostalgic, poetic description of real places in opposition to the depredated and disfigured areas, thus justifying acts of violence against the late 20th century version of the Machine in the Garden. Leo Marx's *The Machine in the Garden: Technology and the Pastoral Ideal in America* (1964), written about the same time the narrative in Abbey's novel takes place, but trying to find a peaceful resolution, back in the 19th century, between the requirements of the Pastoral ideal and the reality of the American Industrial Revolution, can be seen as an ironic counterpart of what is going on in *The Monkey Wrench Gang*.

Deadly serious at times, mock-heroic and parodic at times, engaging with the conventions of the picaresque improved by the western (see Chapter 24: Escape of the Depredator), Edward Abbey's ecodefense, though a period piece of the early days of the radical eco imperative movements, can be re-read now from the perspectives forced upon the world by the somehow unexpected developments brought about by the apocalyptic pages of A.D. 2022.

Works cited

Abbey, Edward. *Desert Solitaire: A Season in the Wilderness.* New York: Ballantine Books, 1971. Print

Abbey, Edward. *The Monkey Wrench Gang.* Tenth Anniversary Edition. Salt Lake City: Dream Garden Press, 1985. Print

Abbey, Edward. *One Life at a Time, Please.* New York: Henry Holt, 1988. Print

Algeo, John and Adele Algeo. "Among the New Words." *American Speech*. Winter 1988. Vol. 63. No. 4: 345–352. Print.

Bly, Robert. *Iron John: A Book About Men.* New York: Addison-Wesley Publishing Company, Inc., 1990.

Garrard, Greg. *Ecocriticism*. London and New York: Routledge, 2004. Print.

Glotfelty, Cherryl. "Introduction: Literary Studies in an Age of Environmental Crisis." *The Ecocriticism Reader: Landmarks in Literary Ecology*. Eds. Cherryl

Glotfelty and Harold Fromm. Athens and London: The University of Georgia Press, 1996. Print.

Knott, John R. "Edward Abbey and the Romance of Wilderness." *Western American Literature*. WINTER 1996. Vol. 30. No. 4: 331–351. Web.

Lloyd, Jeremy. "Rednecks for Wilderness: Earth First! Cofounder Dave Foreman On Being A True Conservative." *The Sun Magazine*. Issue 360 Dec. 2005 Issue 360. Web.

Loadenthal, Michael. "Eco-Terrorism? Countering Dominant Narratives of Securitisation: A Critical, Quantitative History of the Earth Liberation Front (1996–2009)." *Perspectives on Terrorism*. June 2014. Vol. 8. No. 3: 16–50. Web.

Loeffler, Jack. "Edward Abbey, Anarchism and the Environment." *Western American Literature*. SPRING 1993. Vol. 28. No. 1, EDWARD ABBEY: 43–49. Web.

Marx, Leo. *The Machine in the Garden: Technology and the Pastoral Ideal in America*. Oxford and New York: Oxford University Press, 2000. Print.

Mertig, Angela. "From Conservationists to Environmentalists: The American Environmental Movement." *Cultural Dynamics of Climate Change and the Environment in Northern America*. Ed. Berndt Sommer. The Hague: Brill, 2015: 55–76. Print.

Nash, Roderick Frazier. *The Rights of Nature: A History of Environmental Ethics*. Madison: The University of Wisconsin Press, 1989. Print.

Olsen, Brett, J. "Wallace Stegner and the Environmental Ethic: Environmentalism as a Rejection of Western Myth." *Western American Literature*. SUMMER 1994. Vol. 29. No. 2: 123–142. Web.

Phillips, Dana. "Ecocriticism, Literary Theory, and the Truth of Ecology." *New Literary History*. Summer, 1999. Vol. 30, No. 3: 577–602. Web.

Scheese, Don. *Nature Writing: The Pastoral Impulse in America*. New York: Twayne, 1996. Print.

Sessions, George. "The Deep Ecology Movement: A Review." *Environmental Review*. Summer, 1987. Vol. 11. No. 2: 105–125. Web.

The Wilderness Act of 1964 (justice.gov)

Tishler, Nicole A. "Fake Terrorism: Examining Terrorist Groups' Resort to Hoaxing as a Mode of Attack." *Perspectives on Terrorism*. August 2018. Vol. 12. No. 4: 3–13. Web.

Trujillo, Horacio R. "The Radical Environmentalist Movement." *Aptitude for Destruction, Volume 2: Case Studies of Organizational Learning in Five Terrorist Groups*. Ed. Brian A. Jackson. Rand Corporation: 141–175. Web.

Patrycja Pichnicka-Triverdi

Ecological Thinking in Fantastic Literature. Symbolism of New Heroes: A Case Study of 21st Century American Vampire Narrative

Abstract: Vampire narratives are rich in ecological meanings. They revolve around two main issues: the attitude towards non-human beings (the definition and application of such notions as agency, subjectivity and personhood) and the practical relations between man and nature. In opposition to the "bad" vampire, the new vampire figure is not evil and frequently appears as vegetarian. This essay examines the ecological significations in the new Vampire narratives and their relation to posthumanist challenges and prevailing anthropocentric views.

Introduction

Vampire narratives generally revolve around two main issues: the attitude towards non-human beings and the relations between man and nature. These relations – real, projected, imagined or postulated – represented within the narratives, depend on how the non-human subjects (or objects) are defined. Though the topic requires a larger study, it is enough to state here that the insight into the crucial notions of posthumanist theory, in terms of agency, subjectivity or personhood remains rather anthropocentric in most Vampire narratives.

Both subjectivity and agency can be differently defined by diverse theorists in more inclusive definitions (Bruno Latour's conception of agency, for instance) or less inclusive explanations (Markus E. Schlosser), with personhood being the most restricted term, encompassing only some of the agents/subjects. Agency can be assigned or denied to different creatures; when assigned to non-human beings, it is usually condemned as unnatural, evil, abnormal. Such was the case of the traditional Vampire narrative, in which the un-human (thus non-subjective) agency of the vampire was presented only to be condemned and erased. The correct "natural" order was re-stored by the killing of the vampire, an objective carrier of agency in Latourian sense. Traditionally, this kind of agency was thus seen as non-existing or abjective: there were no good objective agents, agency was restricted to subjects, and not even all of them: not to the passive, usually female, victims of the vampire, saved by the brave heroes.

The evolution of the Vampire narrative and the emergence of a new type of hero brought only superficial changes: the non-human beings have been subjectified and personified, that is, humanized.

Vampire attack as nature revenge

The concepts of subjectivity, agency, personhood and humanity are linked to ecological policies that also find their representation in Vampire narratives. Attacks of vampires rising from the dead can be seen as the symbolic revenge of subdued, exploited earth/land. Contemporary chthonic monsters, for instance, appear in recent vampire culture in French-Belgian Franck Richard's movie *La Meute* (2010), which uses Texas for the place of action, and represents the declining postindustrial age. Its ground, soaked with blood, gives birth to monsters, the pack of ghouls indicated by the title, *La meute* (Fr. the *pack*). The ground is feminized even further: Charlotte, the heroine, reads in the newspapers that this ground was raped and that is why it sends its monstrous children. The symbolic rape was done by the intensive exploitation of the mining companies. It was a violence made to the earth – and to the local humans, the laborers, overexploited by the capitalists living in the city. The exploitation of the land and of its people are linked. Resigned bitter people live in dirty grey houses, use old fashioned tools, use wheelbarrows or horse wagons to carry loads.

One representative victim is the La Spack mother (figure of mother-earth) who lost her four sons in the mine. She takes care of the chthonic monsters, faceless (apart from a greedy mouth with fangs), blind humanoid creatures who rise from the ground to claim blood. She, with the help of her last son, Max, who lures victims to his bar, provides the Earth and the monsters with fresh blood. Charlotte is one of her potential victims. But Charlotte seduces Max and, along with a motorcycle gang and an old policeman, the undignified Crew of Light[67], faces the widow and her ghoul monsters. Finally, however, the monsters get their victim. Charlotte is bitten by the ghouls, but the shot shows her face in ecstasy (such ecstasy coming from the brutal bite is quite a common feature of vampire narrative). The last scenes show Charlotte pregnant, fulfilling the role of deceased La Spack, standing at the bar and looking at Max luring the next victims.

---

67 Crew of Light is the name given to traditional brave heroes, vampire hunters, noble opponents of vampires in classical vampire narratives, such as Christopher Craft's "Kiss Me with Those Red Lips: Gender and Inversion in Bram Stoker's Dracula".

Her pregnancy links her to mother-earth while the child she is bearing could be Max's or the ghouls'.

Charlotte herself came from the big city, the power-place dominating the countryside. However, when she appeared in the first scenes of the movie, she was actually running from the city, running from its masculine power symbolized by the motorcycle gang. She came to the land of the exploited vengeful female (woman and earth) and she stayed to face the horrible consequences.

The movie expresses fear of revenge, of reversed colonization. These two features are common for classical vampire narratives. Reversed colonization, for example, is the term applied by Stephen D. Arata to describe Bram Stoker's *Dracula*: here comes a dangerous oriental Other to colonize Great Britain's London, the metropolis of the greatest colonial empire of its time (Arata 85–87). The vampire eventually gets the land (he buys property) and the women of the Western men.

Land, women, chthonic monsters and animals are linked in the colonial-patriarchal imaginary. The taking of the land was marked by subduing the locals, especially women, and animals. Land is usually feminized, and so are the local, non-Western people. Yet, women and the non-Western people are also animalized; they are both discursively and actually linked to animals[68]. Thus, the vampire or the ghoul is not only an othered stranger, it is a radical non-human Other, a predator, hunted down both as an enemy and as an abnormal living thing. Vampires can assume animal shape, can even command forces of nature, such as the wind and the fog. According to Simon Bacon, Dracula or Orlock from Friedrich Murnau 1922 *Nosferatu* movie could represent a whole ecosystem trying to save its integrity (Bacon loc. 363–1178). Such words are put into Dracula's mouth in George DeVein's *The Renfield's Journal*: "I have come from afar to claim my rightful place as the Corrector. Nature has sent me…" (DeVein loc. 3164). The vampire embodied reverse colonization by the non-Western people, as well as the "colonization" on the civilization by nature, a vengeful attack on the human culture. Vampire amounts to the conquest of "dead" passive landscapes, when returning as a living dead with the intention to fight the former, "civilizing" conqueror.

---

68 See, for instance, Michael Spierig and Peter Spierig's movie *Daybreakers* (2009), which illustrates how the consumption of meat and natural products by some means the starvation of others in the opposite relation between the rich blood supply of the Global West/North and the starvation of the Global South.

The vampire is eventually killed and the order is restored – the order which claims itself as natural, consisting in the white (hu)man domination of nature. There are, of course, humans who got involved on the side of the vampire: women who can be easily seduced, nomad people (like the gypsies in *Dracula*), medically diagnosed sick people (like Renfield in *Dracula*). But they are not considered fully human in the narrative and they are meant to perish with the vampire (like the gypsies mentioned above) or to convert back to humanness (like Renfield, who finally decides to help "the brave men" after seeing "their" Mina).

The traditional narrative with the bad vampire killed by the good slayer in the defense of the innocent victim(s) is still extremely popular. In David Slade and Ben Ketai's movie *30 Days of Night* (2007) vampires come with the polar night, freeze and snow, appearing with the storm to destroy human settlements. In Scott Stewart's movie *Priest* (2011) vampires are a hostile, enemy species that humans need to destroy in order to survive. The narrative shows vampires being "a plague on humankind but also the potential (…) eco-warriors that planet Earth desperately needs" (Bacon loc. 65).

Even in the classical model of the narrative, a few ecological postulates may be found. In Franck Richard's *La Meute,* in Francis Lawrence's *I Am Legend* (2007) or in the *Van Helsing* Netflix TV series (2016–2021) the appearance of bad vampires is the result of some ecological/natural catastrophe. So it is in Michael Spierig and Peter Spierig's movie *Daybreakers* (2009), which could be looked at as an anti-capitalist movie. There, the bad vampires are in power. They represent big pharmaceutical companies, exploiting people as sources of food, and exploiting other vampires as their soldiers serving to keep their blood stockage safe. Military measures are needed because of the desperate Third World and/or low-class vampires, who slowly change into zombified monsters due to insufficient blood intake. The structure of the vampire narrative actually serves to reverse the traditional anthropocentric attitude to animal farming: it is white high-class male domination that is showed as disastrous to the global society and to the ecosystem. *Daybreakers* is about oil and pharmaceutical business, and capitalism in general, but it is also about animal rights, as finally vampires start to "trap humans (like animals) to be factory farmed. Humans are heavily sedated and warehoused in cramped factory conditions – the vampire equivalent of veal – where they are machine 'milked' for their blood" (McFarland Taylor 161).

The Strain (2009–2017) narratives go even further: they show people in the butcher houses. However, they do not contain ecological meaning at all; they just show how awful it is when humans are treated as animals. On the same note, *I Am Legend, Daybreakers* or *Van Helsing* also remain ecologically conservative: the remedy to the ecological disaster provoked by humans, who made nature hostile to humans – is still to return to the taming of nature. Nature needs to be under the control of the (hu)man, and if the (hu)man needs to be wiser and less exploitative towards the nature, still his right to rule over nature is not questioned. The right(ful) (hu)man remains the positive hero of such narratives.

Vampire diet

Another ambivalently ecological feature of the vampire has always been his diet. Vampires drink human blood. Therefore, they transgress the border that refers to the very foundation of Western civilization: the culturally constructed difference between human and non-human, or, as it is put within this construction, between humans and animals. Vampires treat humans as food, the treatment which humans reserve only for animals. And vampires need to be killed to address this disturbing transgression. Vampires also need to be killed because they are seen as animals and because, by killing them, men express their right to subdue and consume animals. If in the classic vampire narratives of the 19th century, the Crew of Light is composed of the Vampire Hunters, subsequent narratives that continue this narrative path tend to show the figures of butchers, emotionlessly slaughtering masses of zombified vampires in movies such as *Priest, 30 Days of Night* or Guillermo del Toro and Chuck Hogan's *The Strain Trilogy: The Strain* (2009), *The Fall* (2010), *The Night Eternal* (2011)) and the corresponding TV series (2014–2017).

Finally, vampires need to be killed as Western civilization relies on violence (violence against animals which then permits violence towards, and domination of, other, animalized groups of humans). This human violence is usually hidden in an anthropocentric culture. Vampires unmask this violence. So does *Dracula*'s Renfield, who eats animals just like the Heroes of the story, the only difference being the fact that he kills and consumes animals that are considered to be inedible in Western culture. Because the difference is rather lame, it needs to be medicalized: Renfield is diagnosed with zoophagia and considered a dangerous lunatic.

The important way of masking human violence is by mediating it, by exerting it less directly. While hunters have always used weapons and tools as prothesis mediating their violence, vampires straightforwardly kill with their bodies, using their claws, fangs, and bare hands. This shows an important demarcation between what is seen as animal and what is seen as human. There is also a fascinating logical twist here: the animalistic vampire is in line with the natural world, in which carnivorous animals have bodies equipped to kill their prey. Vampires are in line with nature, but are constructed as unnatural. The very thing that should actually justify the vampire's violence by showing it as necessary for its survival, as a part of its nature, is used as a discursive element of condemnation. At the same time, the mediated violence of men is affirmed as something necessary and good. It is even better visible in Renfield's case, who is judged as a dangerous lunatic when he consumes living creatures which he managed to kill with his bare hands, while solar heroes proudly tell stories about their hunting trips with rifles and guns.

It is also worth noting that in the 20th and 21st century vampire narratives most of the good vampires do not kill with their bodies any more: vampire hunters have swords and guns. This was even made into an issue in the case of the vampire Greyfriar in British Vampire Empire trilogy of Clay and Susan Griffith (*The Greyfriar* (2010), *The Rift Walker* (2011), and *The Kingmakers* (2012)). Greyfriar masters the art of fencing, despising his natural bodily predisposition. In fact, using the sword is uncomfortable to him and far less efficient than if he simply used his claws, but he refuses to do it by principle. Thus, mediated violence, masked, hidden or sublimated, is a mark of civilization, an element of construction of the difference between human and animal. Paradoxically, the mediation of violence is made into a discursive proof of its necessity, while the unmediated character of vampire violence only turns them into a monster.

The vegetarian vampire

The vampire narrative significantly evolved in the 21st century, producing the "Good Guy Vampire" figure[69]. The vampire has not only become a positive hero: the vampire has been promoted to the role of a Hero of Light. Moreover, there are voices[70] that call this new vampire hero a vegetarian, pro-ecological type, opposing cruelty, animal and species chauvinism.[71] The vampire narrative in its new version, with a positive vampire hero, presents a non-human creature

---

69 The term was invented by Margaret Louise Carter in her multiple works on vampires (*Shadow of the Shade: A Survey of Vampirism in Literature* (1975), *Specter or Delusion? The Supernatural in Gothic Fiction* (1987), *Dracula: The Vampire and the Critics* (1988), *The Vampire in Literature: A Critical Bibliography* (1989), *Different Blood: The Vampire as Alien* (2001)).

70 See the articles by George A. Dunn, Jean Kazez or Nicolas Michaud in *Twilight and Philosophy* (2009) and *True Blood and Philosophy* (2010).

71 Species chauvinism is an ideology that claims that the human being is superior to any other creature and never recognizes a fellow being in an animal to be eaten, exploited or kept for pleasure and company.

as worth having (human) rights, as an agent, a subject and a person, as equal to humans. The narrative can be therefore seen as advocacy for other non-human beings. Dale Hudson, for instance, thinks that the plot showing confinement of Damon and Enzo in laboratories in the *Vampire Diaries* serial (2009–2017) can be read as a protest against animal testing (Hudson 209).

Moreover, vampires may be considered a species that is superior to humans and "naturally" feeding on the lower human species. This attitude makes humans realize what it feels like being regarded as a prey (McFarland Taylor 162), which make them eventually empathize with animals that people usually feed on. As McFarland Taylor notes, in *True Blood*, "blood-and-guts aesthetics of drained corpses, half-consumed bodies (…) prompted some viewers to (…) rethink their meat consumption" (159). On the other hand, human audiences get a model hero, the good vampire, who gives up his "natural" food for moral reasons.

Narratives such as *Vampire Diaries,* the *True Blood* 2008–2014 series based on Charlaine Harris's *The Southern Vampire Mysteries* book series (2001–2013) or the *Twilight* book saga (2005–2020), which, in great part, have been turned into movie scripts, introduce vampires who, with diverse means and for diverse reasons, renounce drinking human blood. The vegetarianism of the new vampire hero goes against the Western dominant patriarchal views, in which diet, namely eating meat, has been traditionally seen as a hunter's performance of maleness and manhood. The hero, a male vampire, turned vegetarian, is also close to women. Yet, contrary to the traditional narratives, this is not seen as abject. It is frequently under the female influence that he embraces vegetarianism and feminist ecology.

Ecofeminism uses the link established by patriarchal culture between woman and nature in order to claim the emancipation of both. It also postulates the abolition of the division between the body/matter and the spirit/form, dating back to Aristotle. In this division it was the matter, the body, defined as the female element, that needed to be tamed by spirit, defined as male. The new vampire hero is, as a vampire, an extremely "bodily" male creature. Thus, he transgresses the traditional male-female opposition. In the *True Blood* serial, for example, Sookie gets sexual satisfaction through transgressive intercourses with vampires. In *Twilight*, Bella wants a baby and she gives birth to a new creature, a figure who represents the transformation of the world into a new more symbiotical form of co-existence and co-creation, as Donna Haraway describes in her 2016 study, *Staying with the Trouble: Making Kin in the Chthulucene.*

Meat is performative for Western white male domination. As Homi Bhabha points out in *The Location of Culture* (1994) if the meat supply is limited, the white men mostly deserve to get it; if the meat is plentiful, everyone should eat it, as it is considered (by the white men) as the best nutriment. Yamazaki Masakazu writes about the Westerners seeing the Japanese as feminized, therefore, unable of self-government: "placid vegetarians who lack the determination and enterprising spirit of meat-eating Westerners, who had to hunt their food" (Masakazu 10). The Japanese embraced Western habits and introduced meat in their fish and vegetarian diet as late as the 19th century, in the Meiji era. They coined various expressions, such as *sōshoku danshi* (men-plant-eaters) which is a contemptuous name for men who do not meet the cultural norms of maleness and are, therefore, judged as social pests[72]. The colonial mimicry made the Japanese imitate Western models (including the dietary ones) without ever actually realizing they were becoming westernized. The new type of vampire is actually linked to the Orient, it "implicitly challenges something very iconic and embedded in American culture" (McFarland Taylor 126). This association dates back to the old classic "evil vampire" type, but this time, the vampire has acquired positive features.

Finally, vegetarianism is a metaphor of a broader abstention, which Sarah McFarland Taylor describes as ecopiety:

> The problems of global consumption and its effects on the planet get recast in contemporary vampire narratives as the personal moral and ethical struggle for ecopiety by a virtuous vampire who strives to temper his or her extractive nature (McFarland 126).

Vegetarianism (including the vegetarian vampires as well) challenges capitalism that, like liberal democracy, is strongly linked to meat-eating in the American culture and beyond. McFarland Taylor points at the connection of idioms used to advocate sexual (and procreative) abstinence/temperance/self-restraint and ecological virtue of abstinence from consumerism and consumption, and fossil fuels extraction. Specifically, it is about "the moral act of not doing," recast as active, empowering and even erotic, mirroring "the dual-sided nature of piety (…) which can be associated as much with virtuous renunciation (…) as with virtuous ecstasy" (McFarland Taylor 141, 142). This "longitudinally promised payoff" is "the creation of a more peaceable earth community for all beings" (162).

In green vampire stories, "the conventional association of environmental virtue with piety of self-denial gets re-storied as an *ecstatic* piety of the senses … [which] recasts seemingly dour obligations and collective moral restraint

---

[72] The expression was created by Maki Fukusawa in 2006 (Maki Fukusawa, *Dai-gokai sōshoku danshi,* as quoted in Gabriela Matusiak's article "Wizje erotyki w Japonii" (Matusiak 60).

associated with environmentally virtuous behavior in terms of self-erotic interest" (McFarland Taylor 142). This is also framed as self-enhancing life goals/values, strongly valued in Western, mainly American, culture. Some vampires explicitly promote ecological practices such as recycling (Bill Campton in *True Blood*) while linking it to their practice of restraining from drinking human blood – as the expression of the same attitude.

Vegetarians were already discriminated in the folk lore stories about vampires. The Otherness of Slavic *upiór* relied on a diet, but not necessarily on a bloody one. They frequently drank blood, their own or someone else's, human or animal, before their death (Slavs believed that humans may be destined to become vampires after their death and that some living humans are already vampires). *Upiórs* very often could be recognized by the fact that they did not want to eat a tasty piece of "normal" meat. Some of them, during their lifetime, ate only bread, vegetables and milk (Kolberg 34–35). The suspect Others were meant to be tamed in their graves after they died. And so was the bloody fictional vampire, taken from the Slavic lore and transformed into the main figure of Western modern fiction when Eastern Europe was discovered, or rather created, by the West (as Larry Wolff states in his 1994 book, *Inventing Eastern Europe*). The 19th century fictional vampires were not vegetarian, but they subverted the social order in another way: they were animals trying to eat humans instead of letting themselves be eaten. Therefore, they had to be killed. The new vampires are doubly subversive: they do not deserve to die and are vegetarian, therefore, ecological creatures.

White patriarchal ecology

As Catriona Sandilands notes, ecology was first a tool of racial, class and gender domination (Sandilands web). Primarily the discourse on nature was based on Judeo-Christian theology with its commandment to subdue the earth (Gen 1:28). This ideology was a useful tool in the early colonization of America by Spain and Portugal and it was supported by scientific reasoning during the era of modern colonization. The white heterosexual man had his dominance affirmed by the science he managed and controlled. Animals, females and non-white people were supposed to let themselves be guided by his mind and spirit. Homosexuality was seen in the light of Darwin's theory as something unnatural, as procreation was declared a central element of evolution.

Ecology, as a branch of science, emerged from, as Catriona Sandilands claims, the analysis of the history of the American parks of Yellowstone and Banff. The goal, according to her, was to preserve nature for white men. The parks were created for white middle-class men as places where they could come to express their maleness: have "male" entertainment such as hiking, hunting or other masculine sports. They were places of performance of white homosocial communities, far from women, people of color and the working class. Emancipation of those inferior to them was seen as a threat; it was a threat to the white middle-class heterosexual manhood and therefore to the world in general. "Nature" was an escape from all that was considered "unnatural". This "nature" was a tamed one, organized into a park – thus a cultural, social, human product, a socially constructed version of "nature." "Nature," constructed and organized homosocially (like the Yellowstone or Banff parks or the scouting camps) or heterosexually (like "family" camping or city parks, where spaces were organized to favor heterosexual groups). This reveals that the term of "naturalness" is a purely social invention. If "raw" nature was seen as dangerous, something to be tamed, conquered and colonized by brave heterosexual white men, "naturalness" was part of white heterosexual male performance, while the stigma of "unnatural" was a tool of othering the minorities. This double standard is best visible in the creation of vampires as (representing) forces of nature and yet being "unnatural" and "abnormal."

The cases of homosexuality among wild animals were either ignored, or explained with reference to heterosexual procreation, or put down to some ecological disaster, according to Sandilands' article "Unnatural Passions?: Notes Toward a Queer Ecology" (2005). Much in the same way, sexual preferences, orientations, identities and expressions, judged nonnormative, were classified as illnesses and deviations. Some even accused the environment and pleaded for the treatment of "natural" environments. The rise of unnatural creatures of nature, who strongly embody all the transgressive forms of sexuality, is still linked to the "unnaturalness" of the environment and of human actions in narratives such as the *Hemlock Grove* series (2013–2015) or the *Penny Dreadful* series (2014–2016).

The first ecological discourses sounding the alarm against the destruction of the planet were made from the dominating group concerned to preserve the environment for themselves and in the form they wanted it to be. They were closely linked to the countryside nostalgia discourse, with its sexual, gender and national aspects (countryside as a span of "healthy," "natural" behavior, traditional patriarchal relationships and the preservation of national values). Regarding the countryside, the discourse of the dominating center is similar to the colonial discourse: it is either a discourse of open discrimination (the modern city as superior to the backward village), or a nostalgia discourse which wants the countryside to freeze in time, to remain a "traditional," idealized, idyllically "natural" village, regardless the actual needs of its inhabitants.

A new heterosexual white good vampire as a new version of the white male hero may be seen not as a transgressive figure, as argued earlier, but as a conservative one. He may represent the traditional, conservative ideals of patriarchy, of the middle-upper class and racial domination just as the classic white hero-hunter did in the 19th century. Edward (*Twilight*), Stephan (*Vampire Diaries*) or Bill (*True Blood*) treat their "vegetarianism" as a kind of restriction that is to prove their maleness by showing how they can control their bodies by the strength of their minds. "Mind over Matter" is titled one of the chapters of Stephanie Meyer's 2005 *Twilight* novel. These figures reluctantly agree to build a relationship with their beloved women because they are not sure of their procreating possibilities and because they want – just as traditional hero-hunters- to keep the woman pure and able of exercising her motherhood. They do it despite the woman's clearly announced wishes. Eventually, the "vegetarianism" of the good vampire consists most frequently in killing animals and drinking their blood.

The fake ecology of the good vampire

Good vampire figures in fact lose their ties with nature: they do not change into animals, plants or into an atmospheric phenomenon, such as mist. They usually lose their connection with (their) earth. 19th century vampires, such as Dracula or Carmilla from Joseph Sheridan Le Fanu' s tale, were strongly connected to the earth: Dracula needed to carry some of his motherland soil with him, Carmilla needed to return to her grave. Those habits do not exist in most of the contemporary narratives about good vampires.

For Nick Groom, the vampire has always been a figure that subverted the anthropocentric point of view (Groom 128–129). For that reason in classic versions of the vampire narratives vampires were supposed to be killed by human hunters. In the "Good Guy Vampire" narratives vampires are not supposed to be killed, but vampire-ness is erased by means of neutralization. When a vampire becomes the hero, he becomes so similar to his human predecessors, the Crew of Light, that the difference is actually liquidated. He is humanized. The vampire's recognition as a person (subject and agent) is not really a recognition of a non-human being as such; on the contrary, this recognition depends on the humanization degree of the being.

Finally, the animal-eating "vegetarianism" of the good vampire is anthropocentric *par excellence* (Hudson 168). When the vampire Edward Cullen from *Twilight* recounts his hunting trips, he poses as a 19th century colonizer, he chooses only the dangerous predators that he fights without any weapon. In fact, even the strongest predator does not stand a chance against a vampire, so the duel can hardly be called equal. The teasing and fighting the animal that is meant to die anyway is rather cruel, symbolically expressing a very colonialist desire to conquer nature. However, brave, male Cullens never hunt other animals, as there is no fun in that. When, especially in the *Twilight* movies, Bella's vegetarianism is paired with Edward's, "[i]n the process, non-human animals are rendered figurative tofu and their value as living, breathing, feeling creatures largely denied" (McFarland Taylor 144).

Another version of vampire "vegetarianism," the drinking of artificial blood, is closer to actual vegetarianism. However, the very fact that this blood is artificial, it is ecologically problematic. In Charlaine Harris's book, artificial blood is produced by a few different brands. In the *True Blood* serial there is only one brand producing it. Such blood is a capitalist product, a market substitute of sacrum by a commodity (Chaplin loc 941–951), an element of the system of power, in which the work of scientists is a tool (or the core) of domination. Artificial blood is thus related to the exploitative system of production. Inclusion of the Other depends on their assimilation and "vegetarianism" is a lifestyle, an image-building, rather than a moral attitude. It is such in most of "vegetarian" vampire narratives. In the *Twilight* saga, the actual vegetarianism of Bella is meant to parallel the vampire "vegetarianism" of Edward (who hunts animals). Thus, the actual object of "vegetarian" care will matter less; what matters here is the self-construction of the "vegetarian." Bill's recycling and blood abstention are really linked only by the subject that is a carrier of them. Vegetarianism is yet another commodity to consume while self-restraint, self-control and discipline are regarded as traditional elements of masculinist discourse.

According to McFarland Taylor, ecopiety is "a contemporary practice of environmental virtue, through daily, voluntary works of duty and obligation" and it evokes "a harmonic model of proper relations cultivated between humans and more-than-human earth" (McFarland Taylor 3). Most of the vampire narratives promote personal, individual, private acts of ecopiety, in which everyone does their bit (McFarland Taylor 3). Firstly, this is dangerous as it creates the impression that there is no need to make any fundamental structural changes. It constructs an "imagined moral economy in which tiny acts of voluntary personal piety, such as recycling a coffee cup (…) can be exchanged as an offset to justify the continuance of current consumption patterns and volume" (4). We know those practices are not sufficient. Secondly, this stance perpetuates the existing relations of power – McFarland focuses on the economic power, but this applies to political, symbolical powers as well: "Ecopiety, as represented in and through contemporary popular culture, is about cultivating a proper and (…) responsible connection between individual citizen consumers and the more-than-human earth" (4).

Thirdly, as the author notices, this individual ethos is deeply capitalist at the core. "[G]reen consumer marketing is quick to offer a template" (McFarland Taylor 5) of what one should do: global capitalism and market ideology convince it "requires the performance of a correlative 'consumopiety' or 'virtuous consumption'" (4). Therefore, such practices can sustain the very system that is responsible for the ecological draining of the planet. Indeed, in such narratives as *True Blood,* vampires only get ecological by a proper ecological product (bottled *True Blood*). It is also worth remembering that almost all of those ecological narratives are meant to fund a franchise, with multitudes of gadgets and products to sell in a sort of festival of extractivism, according to the same author:

> Green vampires embody environmental moral critiques of *extractive capitalism* and its resource-sucking planetary consequences. And yet, ironically (...) they fuel this moral engagement through the very extractives consumerism they purport to trouble (McFarland Taylor 124–125).

The franchises incite consumerist desire by practices external to narratives and internal to them: they are "tapping into audience's own erotic desires to be all-consumed" (McFarland Taylor 144) and to consume. The gore aesthetics itself is being presented as erotically alluring. Therefore, the potential anti-meat-eating visual dimension is again double-sided.

Fourthly, the narratives promote individualism as the fundamental human attitude, while it is actually a modern Western capitalist one. Of course, not all of the narratives promote individual action only. But even if narratives call for collective action, they are mainstream, so they call to modify the system rather than to actually abolish, radically transform it (McFarland Taylor 160).

Finally, the "vegetarian" diet is presented in categories of restriction and lack: both tofu for humans and animal and artificial blood for vampires cannot satisfy, as stated by vampire Edward Cullen in *Twilight*. They cannot give full strength either – which is in fact a core element of carnist discourse argumentation. It is also a very ethnocentric, Western white discourse. The meat is always missing, the vegetarian diet is conceived as a diet without meat, referring to Western traditional middle-upper class diet, while there are cultures (Indian, Nepali) in which the vegetarian diet is a point of reference and meat in general is seen as a weird and abject item. Viewing vegetarians as weaker can also be a heritage of the Oriental discourse.

Conclusion: Ambivalent vampire – The only non-anthropocentric one?

An ambivalent tragic vampire, such as the one introduced by Anne Rice in her first vampire novel *Interview with the Vampire* (1976) could be considered the most ecological. It was a disturbing figure of the (post)modern condition. By its hunger for blood, who held an obviously sexual character, it metonymically evoked desire of unlimited consumption in general. This (self) destructive desire was critically exposed, yet the non-human vampire was not condemned. In 21st century vampire narratives there is such a vampire hero in *Hemlock Grove,* an heir and a product of medical industry, fighting his nature by finally destroying all that he loves in the consuming frenzy. Abby from Matt Reeves *Let Me In* (2010) is another ambivalent vampire, a product of the internal hidden violence of contemporary society. On the other hand, even this kind of vampires are deeply human/anthropocentric: they are humans with some ailments or humans turned into vampires. Their nature is defined by the transgression of humanity – which makes humanity a permanent point of reference.

The figure of the radically different ambivalent vampire has however appeared in Suzy McKee Charnas *The Vampire Tapestry* (1980). The vampire here is a non-human distinct species, and yet he is a person. He hunts humans because he is a predator and they are his natural prey. He only looks like them because he used mimicry as a biological hunting facility (it is a means frequently used by predatory animals). His psyche is not similar to the human one, rather to the feline one (Carter 358–360). He does not want to integrate into human society and, being outside it, he cannot be judged according to human morality which is nothing else, but a social product of the humans. Thus, his killings are not crimes, as he does not belong to the social/moral order, but to the biological one. However, while remaining outside society, he is no less a person and the narrative condemns those who caught him, held in the cage and experimented on him. Precisely because he belongs to the biological order of nature in which all animal (including human) beings are equal, he cannot be inscribed into the human social order of evaluation of species which hierarchizes beings, ranking humans at the top and giving them all the rights over other animal beings.

In other narratives which present vampires as separate species, vampires are nonetheless usually evil: in *Priest* their lives and highly developed social structure do not interest any protagonist and are not meant to interest the audiences. In the 2019 *V-Wars* serial vampires see the proclamation of themselves as a separate species as an identity strategy used to seek emancipation. Nevertheless, they need to conform to the rules of the human society in order to avoid being killed. In Peter Watts's *Firefalls* novels *Blindsight* (2006) and *Echopraxia* (2014), vampires are a distinct species and are not definitely bad. Yet they are presented as lacking something (human consciousness) and thus their dominance on the planet would be deeply philosophically tragic. The fact that some vampires know it themselves – and thus willingly sacrifice themselves to save humans as the only fully conscious beings of the universe, the only real persons, marks a trend back from previous posthumanist attempts to go beyond the prevailing anthropocentric views of most contemporary culture.

References

Adams, Carol J. *The Sexual Politics of Meat: A Feminist-Vegetarian Critical Theory. 20th Anniversary Edition*. New York: The Continuum International Publishing Group, 2010. E-book.

Arata, Stephen D. "The Occidental Tourist: Dracula and the Anxiety of Reverse Colonisation." *Victorian studies* 33(1990): 621–645. Web.

Bacon, Simon. *Eco-Vampires: The Undead and the Environment*. Jefferson, USA: McFarland, 2020. E-book.

Bhabha, Homi K. *The Location of Culture*. London: Routledge, 1994. Print.

Carter, Margaret L. "The Vampire Tapestry." In: Jane Pulliam "Twilight." *Encyclopedia of the Vampire: The Living Dead in Myth, Legend, and Popular Culture*. Ed. Sunand T. Joshi. Santa Barbara, Denver, Oxford: Greenwood, 2011. E-book.

Chaplin, Susan. *The Postmillennial Vampire. Power, Sacrifice and Simulation in True Blood, Twilight and Other Contemporary Narratives*. Leeds: Palgrave Macmillan, 2017. Print.

Craft, Christopher. "Kiss Me with Those Red Lips: Gender and Inversion in Bram Stoker's Dracula." *Representations* 8: 107–133. Web.

DeVein, George. *The Renfield's Journal*. Scotts Valley: CreateSpace Independent Publishing Platform, 2018. E-book.

del Toro, Guillermo and Chuck Hogan. *The Fall*. New York: William Morrow, 2010. E-book. Web.

del Toro, Guillermo and Chuck Hogan. *The Night Eternal*. New York: William Morrow, 2011. E-book. Web.

del Toro, Guillermo and Chuck Hogan. *The Strain*. New York: William Morrow, 2009. E-book. Web.

Dunn, George A., Rebecca Housel, and William Irwin (eds.). *True Blood and Philosophy: We Want to Think Bad Things with You*. Hoboken, NJ: Wiley, 2011. Print.

Griffith Clay, Griffith, Susan. *The Greyfriar*. New York: Pyr, 2010. E-book. Web.
Griffith Clay, Griffith, Susan. *The Kingmakers*. New York: Pyr. 2012. E-book. Web.
Griffith Clay, Griffith, Susan. *The Rift Walker*. New York: Pyr. 2011. E-book. Web.
Groom, Nick. *The Vampire: A New History*. New Haven and London: Yale University Press, 2018. Print.
Haraway, Donna. *Staying with the Trouble: Making Kin in the Chthulucene*. Durham: Duke University Press, 2016. Print.
Housel, Rebecca, Jeremy Wisniewski, and William Irwin (eds.). *Twilight and Philosophy: Vampires, Vegetarians, and the Pursuit of Immortality*. Hoboken, NJ: Wiley, 2009. Print.
Hudson, Dale M. *Vampires, Race, and Transnational Hollywoods*. Edinburgh: Edinburgh University Press, 2017. Print.
Kolberg, Oskar. *Sanockie-Krośnieńskie*, cz. 3. Wrocław: Polskie Towarzystwo Muzyczne, 1973. Print.
Latour, Bruno. "Agency at the Time of the Anthropocene." *New Literary History* 45 (2014): 1–18. Web.
Matusiak, Gabriela. "Wizje erotyki w Japonii." *Cool Japan. Autoprezentacja Japonii w popkulturze*. Ed. Małgorzata Gotowska and A. Wosińska. Bydgoszcz: Kirin, 2017. Print.
McKee Charnas, Suzy. *The Vampire Tapestry*. New York: Orb Books, 2008. Print.
Meehl, Brian. *Suck It Up*. New York: Delacorte, 2012. Print.
Meyer, Stephanie. *Breaking Down*. New York, Boston: Little, Brown and Company, 2008. Print.
Meyer, Stephanie. *Eclipse*. New York, Boston: Little, Brown and Company, 2009. E-book. Web.
Meyer, Stephanie. *Midnight Sun*. New York, Boston: Little, Brown and Company, 2020. Print.
Meyer, Stephanie. *New Moon*. New York, Boston: Little, Brown and Company, 2008. E-book. Web.
Meyer, Stephanie. *Twilight. The Tenth Anniversary Edition*. New York and Boston: Little, Brown and Company, 2015. E-book. Web.
Masakazu, Yamazaki. *Individualism and the Japanese: An Alternative Approach to Cultural Comparison*. Trans. Barbara Sugihara. Tokyo: Japan Echo Inc., 2000. Web.
McFarland Taylor, Sarah. *Ecopiety: Green Media and the Dilemma of Environmental Virtue*. New York: New York University Press, 2019. Print.
Rice, Anne. *Interview With the Vampire*. New York: Ballantine Books, 1991. Print.

The Holy Bible, English Standard Version. ESV® Text Edition: 2016. Copyright © 2001 by Crossway Bibles. Web.

Sandilands, Catriona. "Unnatural Passions?: Notes Toward a Queer Ecology." *Invisible Culture. An Electronic Journal for Visual Culture* 9 (2005): Web.

Schlosser, Markus E. "Agency, Ownership, and the Standard Theory." *New Waves in Philosophy of Action*. Jesus Aguilar, A. Buckareff, and K. Frankish (eds.). Houndmills, Basingstoke: Palgrave Macmillan, 2011. Print.

Wolff, Larry. *Inventing Eastern Europe: The Map of Civilization on the Mind of the Enlightenment*. Stanford: Stanford University Press, 1994. Print.

Movies:

Condon, Bill. *The Twilight Saga: Breaking Dawn – Part 1*. 2011.

Condon, Bill. *The Twilight Saga: Breaking Dawn – Part 2*. 2012.

Hardwicke, Catherine. *Twilight*. 2008.

Lawrence, Francis. *I Am Legend*. 2007.

Reeves, Matt. *Let Me In*. 2010.

Richard, Frank. *La Meute*. 2010.

Slade, David, and Ben Ketai. *30 Days of Night*. 2007.

Slade, David. *The Twilight Saga: Eclipse*. 2010.

Spierig, Michael, and Peter Spierig. *Daybreakers*. 2009.

Scott Stewart, Scott. *Priest*. 2011.

Weitz, Chris. *The Twilight Saga: New Moon*. 2009.

Serials:

Hemlock Grove. Netflix. 2013–2015.

Penny Dreadful. Showtime. 2014–2016.

The Strain. FX. 2014–2017.

True Blood. HBO. 2008–2014.

Vampire Diaries. The CW. 2009–2017.

Van Helsing. Netflix. 2016–2021.

V-Wars. Netflix. 2019.

The Authors

Anouk Aerni is a PhD student at the University of Basel after she completed an MA in Literary studies at the same university. She is a recipient of the start-up grant from the Doktoratsprogramm Literaturwissenschaft of the University of Basel, a member of the Eikones Graduate School. She is currently doing research on the relationship between humans and nature in American postmodern postwar literature and writing her dissertation under the supervision of Prof. Dr. Philipp Schweighauser.

Loredana Bercuci is assistant professor at the Department of Modern Languages of West University of Timișoara, where she teaches American cultural studies, academic writing and applied linguistics. She holds a PhD in American Studies, with a focus on trauma and transmedia storytelling. Her research interests include trauma studies, adaptation studies, the representation of race in American popular culture, and critical theory.

Pauline Boisgerault is a PhD student at the University of Rennes 2, in France. Her research focuses on space in North-American literature and particularly in the works of Frank Norris, Gary Snyder and William Giraldi.

Adina Ciugureanu is Emerita Professor of English and American Literature at *Ovidius* University Constanta, Romania. She was the Dean of the Faculty of Letters between 2004 and 2012 and Director of the Institute for Doctoral Studies at *Ovidius* University between 2016 and 2019. She is currently the Romanian Association for American Studies (RAAS) and served as Treasurer of the European Association for American Studies (EAAS) between 2008 and 2016. Her research area ranges from British and American culture to urban studies, ecocriticism and geocriticism. She has supervised PhD research in the above-mentioned subjects. She has had numerous research grants at prestigious universities and two Fulbright grants at UNLV, Nevada (2001–2002) and at UCSB, California (2016–2017). Her publications include 6 books (among which *Post-War Anxieties* (2006), *Modernism and the Idea of Modernity* (2004), *The Boomerang Effect* (2008, 2002), *High Modernist Poetic Discourse* (1997), over 50 articles and volume chapters and 8 co-edited volumes based on the conferences on American studies she has organized.

Daniel Clinci is an Assistant Professor at *Ovidius* University of Constanța and holds a PhD from the University of Bucharest. During his doctoral studies, he had the opportunity to do research on contemporary philosophy and media studies at the University of Amsterdam. He is also co-editor of *Post/h/um. Jurnal de studii /post/umaniste*, an independent journal dedicated to bringing new critical theories and concepts into Romanian through the translation of recent academic articles by researchers worldwide. He published *Avangardă și experiment. De la estetica negativă la cultura postmodernă*, București, 2014.

Alina Cojocaru is an Assistant Professor at *Ovidius* University, Constanța from which she also obtained her PhD in English literature. Her research interests include geocriticism, memory studies, spatial literary studies and literary theory. She has written articles on topics related to the above-mentioned areas of study in books and journals indexed in international citation databases and has participated in national and international conferences in the UK, Germany, Austria and Romania. She published the study *Geographies of Memory and Postwar Urban Regeneration (London as Palimpsest)* published by Cambridge Scholars, 2022.

Andreea Cosma is holds a PhD from *Ovidius* University Constanta. She is a graduate teaching assistant and has conducted research in geocritical readings of, and activist manifestations in, the literary works of the Beat writers under the supervision of Professor Adina Ciugureanu. Shee has a BA in American Studies and an MA in Anglo-American Studies, both at *Ovidius* University. She is also an alumna of the SUSI program, Global Social Entrepreneurship Institute for European Leaders, Indiana University, USA, in 2015.

Philip John Davies is Emeritus Professor of American Studies, De Montfort University, Leicester. He is best known for his expertise on US election campaigns and he has also published work on many other aspects of US politics, society and culture. He has authored and edited around thirty books and special journal issues and dozens of articles and book chapters, often in partnership with respected colleagues, on such subjects as US politics and film, politics and science fiction, the American city, the US Constitution, political marketing, the US presidency, and on many aspects of US elections. He has written about the US religious groups, the Shakers, and published several articles on Latvian electoral politics in the early post-Soviet years. He is co-author with Robert McKeever of the successful textbook, *Politics USA* (Routledge, 2012). He has served as President of the European Association for American Studies, as Chair of the American Politics Group of the UK, Chair of the UK Council of Area Studies

Associations and Chair of the British Association for American Studies. He has taught at the Universities of Manchester, Coventry and at London University's School of Advanced Study in the UK, and at the Universities of Maryland and Massachusetts, Creighton University and Wartburg College in the USA. He also served as Director of the David and Mary Eccles Centre for American Studies at the British Library in London, where he managed a large program of lectures, conferences and fellowships.

Carla Francellini is a Senior Researcher in Anglo-American literature at the University of Siena. Her interests range from Translation and Adaptation Studies, Steinbeck's Novels, and Melville Studies to Modern and Postmodern American Novel and Italian American contemporary literature. She edited *Uè Paisà* (2012), *Women in Translation. Donne in Traduzione* (2014), *Miraggi italiani. Tony Ardizzone, Adria Bernardi, Paola Corso, Kenny Marotta* (2019), and co-edited *Re-Mapping Italian America. Places, Cultures, Identities* (2018). She is the author of several articles on Bloom, Coetzee, Woolf, Steinbeck, Melville, Plath, M.Mazziotti Gillan, and the monograph *Visible/Invisible. Incursioni nella narrativa italiana americana contemporanea* (2018).

Oana-Celia Gheorghiu holds a PhD in British and American Literature (2016) and is an Assistant Professor at the Faculty of Letters, "Dunarea de Jos" University of Galați where she teaches British and American Culture and Civilization and Cultural Representations in the Anglophone Space. She is the author of *British and American Representations of 9/11. Literature, Politics and the Media* (2018, Palgrave Macmillan). Her research interests are: political fiction, 21st-century literature, reality and fiction.

Ludmila Martanovschi is Associate Professor at the Faculty of Letters, *Ovidius* University, Constanta, Romania. Her areas of expertise include Multi-Ethnic Literatures and Contemporary American Drama. She has published two books and has contributed to various collections of essays such as *Migration, Diaspora, Exile. Narratives of Affiliation and Escape* edited by Daniel Stein et al. (Lexington Books, 2020). She has served as the Secretary of the Society for Multi-Ethnic Studies: Europe and the Americas (MESEA) since 2016 and has been a member of the European Association for American Studies and the European Society for the Study of English since 2000.

Roger L. Nichols is Emeritus Professor of History & Affiliate Prof. American Indian Studies at the University of Arizona. He holds a PhD in US history from the

University of Wisconsin, Madison. He is the author of *Tombstone, Deadwood, and Dodge City:* Oklahoma, 2018, *Indians in the U.S. and Canada,* U. Nebraska, 2018, *Black Hawk and the Warrior's Path,* Wiley, 2017, *Natives and Strangers: A History of Ethnic Americans,* Oxford, 6 eds. 2015-1979, *American Indians in U.S. History.* U. Oklahoma, 2014, 2003, *Warrior Nations: The United States and Indian Peoples.* U. Oklahoma, 2013, among others. He has been awarded many research grants among which three Fulbright Scholar grants.

Dragoș Osoianu is an English teacher in the pre-university system. He holds a BA and an MA degrees in Filology and Theology and a PhD in English and American Literature with the thesis *Urban Ecocriticism and T. S. Eliot's The Waste Land* (University of Craiova Press: 2018). His major research interests include Ecocriticism, Ecosophy, Dark Ecology, Ecotheology and Ecopedagogy. He has also published ten articles related to Ecocriticism and Ecopedagogy.

Patrycja Pichnicka-Triverdi is a Graduate of Cultural Studies and a PhD student at the Institute of Polish Culture at the University of Warsaw. She has published in *Polish Journal of Political Science* and *Kultura popularna*. Her interests concern postcolonial studies, intersectional theory, social philosophy and popular culture. She is currently writing her dissertation on representations of 'Otherness' in Western and non-Western countries.

Michaela Praisler is a tenured Professor of Anglophone Literatures at the Faculty of Letters, "Dunarea de Jos" University of Galați, where she has been teaching since 1990. She has authored more than 80 books and articles, has led international projects in the area of Cultural Studies, and has supervised many postgraduates on their path to PhD. Her research interests concern Modernism, Postmodernism, literary theory, feminist criticism, metafiction, and Film Studies. Michaela Praisler is general editor of the journal *Cultural Intertexts* (est. 2014).

Olga Thierbach-McLean is an independent researcher, author, and literary translator. After studying North American literature, Russian literature, and musicology at the University of Hamburg and UC Berkeley, she earned her doctorate in American Studies at UHH. She is the author of the book *Emersonian Nation.* She is also a contributor to *Amerikastudien/American Studies Journal, The Irish Journal of American Studies, The European Journal of American Studies, Swiss Papers in English Language and Literature,* and *U.S. Studies Online.* Her main research interests are in US political culture, American Transcendentalism, the intellectual history of liberalism, and dystopian fiction. Currently, her projects are focused on the significance of race in the cyberpunk genre as well as on reinterpretations of traditional individualist tenets in contemporary US cinema.

Florian Andrei Vlad is an Associate-Professor of American Studies at the Faculty of Letters, *Ovidius* University, Constanța. He holds an M.A. in American Studies from the University of Heidelberg and a PhD from *Ovidius* University, Constanta. His recent publications include *Space, Place, Narrative in John P. Quinn's Poetry* (2020), *Literary Lights and Shadows in the Post-9/11 Age: Literature, Trauma, Geopolitics* (2021), and *Challenging Identities: From Theory to Literary Discourse* (2022).

Eduard Vlad is Professor of English and American Culture and a PhD advisor in the Doctoral School for the Humanities at *Ovidius* University, Constanta. A former president of the Romanian Association for American Studies, he has shown increasing interest in areas of research embedding the literary text in wider interdisciplinary research areas. Moving from volumes on British poetry and fiction, including his PhD thesis on Philip Larkin his more recent interest in American studies can be shown in such volumes as *Globalization, Geopolitics, and the US*, *Cultural Studies: Archaeologies, Genealogies, Discontents,* and his *Dictionar polemic de cultura americana,* as well in his co-editing *Ideology, Identity, and the US: Crossroads, Freeways, Collisions* and *National and Transnational Challenges to the American Imaginary.*

www.ingramcontent.com/pod-product-compliance
Ingram Content Group UK Ltd.
Pitfield, Milton Keynes, MK11 3LW, UK
UKHW041902230426
12049UKWH00002B/19